AT HOME IN THE REAL WORLD

Suzy Meyer

First Edition

ISBN 978-1-54391-433-7

Published by Calamity May, LLC

P.O. BOX 99751

Pittsburgh, PA 15233

Printed in the United States

Covert Art: Stoneypath Tower, East Lothian, Scotland. After a painting by Andrew Spratt. Used with Permission.

Cover Design: Sharon West Design Consulting.

DISCLAIMER

The materials within this book represent the findings of the author. Neither Suzy Meyer, nor Calamity May, LLC is to be held responsible, accountable, or liable for anyone's interpretations, personal experiences, actions, or outcomes.

DEDICATION

To those who've endured violence,
and those who haven't.

CONTENTS

About This Book
Introduction

1 PART I. AWARENESS

2 PART II. PREVENTION

3

PART III. DEFENSE

R

REFERENCES

Appendices

A: Crimes defined, FBI

B: Newspaper Headlines: Underreported Crime

C: Elementary Schools to Universities with Potential Violations for Mishandling Sexual Violence

D: 5 Supreme Court Cases: Police Have No Duty to Protect (Myth 1)

E: The Narrative of Castle Rock v. Gonzales (Myth 2)

F. Considerations for New Doors

Footnotes
Bibliography

ACKNOWLEDGMENTS

Just as a Queen of Denial is not unmade in a day, a book like this isn't written in a vacuum. I have many to thank for encouragement, training, feedback, support, and real-world expertise. My list of thanks and gratitude begins at a kitchen table and ends where, I know not. You are: Terry Riebling, Ed Siemons, Brian Keith, retired law enforcemnt officer, (for the initial book experiment and ongoing valuable mentorship); Laurence O'Neal Suarez, strategist and editor, (recognized the value of daylighting taboo topics, and focused this author's energy); Keith Bushy & the Bulldog Jeet Kune Do team; Retired SOCOMM Major Cyrano/ex-operative, (turned naivety into grit); Keith Attwood, goldsmith, (unwavering encourgagement, spirited conversations, Tai Chi instruction); Phyllis Tustin, Crosswind K9, (genteel badassery); Irma, Connie, Kurt, Pam, Kathy, Jim, Lloyd, Diana, Sherri, Miriam, Nanette, Amanda, Alexander....(Round 1 beta readers); Attorneys Patrick Tomassey and Philip Irani, (counsel); Tom Gigliotti, photographer (headshot); Ab Meyer (self-defense model); Kurt Hess (illustrations); Sharon West, Sharon West Design Consulting (branding, website, and book layout– how she did it I'll never know); Matt Keith, Klint Macro, Woj (firearms training); Jim Hamel, PhD., aka Dr. Revolver ("Make your first shot count"); Paul Chandler, Altra Firearms, (patient, helpful, and respectful to newbie gun buyer); Steve Schwarzer, mastersmith, (blade basics); Choppo, sporting clays enthusiast, (thanks for the moving targets!); all the guys at CDSA Wednesday night indoor shoot; Stephen Barron, Chad Allen, Brian Keith, Molly, Kathy Newman, Matt Kieth, Klint, Macro...(Round 2 pro level readers); a friend (who got me to the last lap); and finally, my family. To those not named, forgive me. You know, I know.

This book exists because of my husband, Joe. He allowed me the time to research, write, and train. Together, we learned strength training and Bulldog. I'm pretty sure I've tested every fiber of his person, which he endured for the love of the women in his life.

ABOUT THIS BOOK

With modest depth and breadth, this book shows you how to greatly reduce the potential for break-ins, and how to defend yourself should one occur when you're home. To help you develop a balanced, proactive mindset and learn effective strategies, I build upon three well-established principles: Awareness, Prevention, and Defense.

PART I. AWARENESS Update your perception of the world around you: know how it really works.

First, learn the facts: research what crimes against people and property exist where you live; familiarize yourself with American law; and, understand the real limits of police protection.

Second, understand how you and your home may be perceived by predators; learn to assess your dwelling's vulnerabilities and overcome them.

Third, recognize and cultivate two fantastic tools you were born with: intuition and survival instincts. While you're of sound mind and body, be the one responsible for you personal safety. Don't assume that your partner, dog, local police, or government 'should' or will do it for you. Remember,

Never Abdicate Your Personal Safety!

PART II. PREVENTION Secure your residence! Deter would-be intruders from getting into your home. If you're home during a break-in, much is at stake. By the end of this section, I hope to have successfully conveyed why you need a cohesive, integrated prevention system that fits your lifestyle.

This book shows you 28 tools and best practices to help you build a home prevention system based on your needs, budget, and abilities. Some are free, some are retail products, some are do-it-yourself (DIY) projects. At the end of this section, I share with you the preventive measures we implemented in our home over the last three years, and our post-evaluation including costs and ease of installation. Why all this? Because,

You Want Prevention More Than Self Defense!

PART III. DEFENSE If your preventive measures fail, (and they can), you need an immediate response plan. With proper mindset and tools, you can stop an intruder in his tracks! Think through scenarios that align realistic threats with your abilities to respond. Not sure how to think about it? *Neither was I!* This section will help you uncover what you are willing and able to do.

"What Am I Willing To Do?" is a gritty exercise I developed. By going through it, I learned a lot about myself and what I'm made of. Try it, I bet you'll surprise yourself, too. You'll also find encouragement to cultivate Primal You (your survival instincts), regain health, increase core body strength, and fine-tune ancient skill sets—no matter your age.

As for self defense, this former Queen of Denial turned Home Defender explores unarmed self-defense, improvised weapons and non-lethal pepper spray. I also share my learning curves and considerations for using firearms for defense in the home. Difficult questions are posed. Proper protocols are shared and encouraged.

The book concludes with legal terms we all need to know. In the end,

Have A Plan! And A Backup Plan!

A NOTE ON FIREARMS FOR HOME DEFENSE

I have come to agree with millions of women and men who already own a handgun and/or shotgun for self defense in the home: to own a gun, to safely handle and practice with it, to know how and when to use it, is the most effective means within a small space and a few seconds, to save your life. Period. I underscore what has been said by many self–defense educators before me: **A firearm for home and self defense is a tool of last resort. Firearms are no substitute for not having preventive measures to keep bad people out of your house.** Read this book, adopt your own prevention system to secure your yard, entryways, doors, and windows.

The more options you prepare for, practice, and have at-the-ready, the greater your success in preventing intrusions and countering dangerous situations. This isn't paranoia, it's an informed response in an uncertain world.

INTRODUCTION

This book is meant to help all women and men living under American law, no matter the neighborhood you live in or the size of your house; what car you drive or what bus you take; your age, politics, or religion. This book squares off with subtle fears and denial that many of us have about personal safety and home security.

What you are holding in your hands is a record of my journey from Queen of Denial to a more fully aware citizen and confident human being—the emergence of Calamity May.

In Fall 2011, I lost my job and started working from home. As a landscape architect and garden designer, I expanded our garden and learned to grow food. I was outside a lot and that's when I heard the gunfire, usually in the day, sometimes at night. It was common to hear two or three shots in a row. Sometimes there were five, or just one. With every shot I felt alarmed, afraid, and pissed off. Each blast was an affront to my well being, a threat to my safety. I often reported them to the police.

The Summer of 2012 it escalated. I heard approximately 40-50 separate incidents emanating from the valley below our house. One day in September, while tying up a tomato vine, a police bullhorn, 500 feet away blared, "Come out with your hands up!" I froze; dropped my work; held my breath; and grew large ears. The command was repeated. Four methodic shots were fired. Then one more. Then nothing….except disbelief.

I was shocked; ready to move from the house we bought and renovated three years earlier. I was mentally packing my bags when my husband walked in from work. "Joe, we gotta move." As he listened, I could tell he was tallying the work we had put into this 90-year old house, the money sunk into it for structural soundness and energy efficiency. Then he brought up the obvious—that we didn't have enough money to make a good move. "And, if we did have more money Suzy, where exactly would we go that's safer? Besides, this is our home, dammit." Uh oh. My ears were hearing Stand Your Ground talk.

Joe grew up on a farm in southwest Pennsylvania. His family, like many who work the land, are independent, resilient, hard working, and proud of their labors. They don't move when things get hinky. These people put down taproots and are hard to transplant.

The next morning I pored over the City of Pittsburgh's annual police reports and maps to learn that our neighborhood was on the edge of a statistical zone of crime and violence. Gunfire was attributed to drugs, money, and retaliation. Our neighborhood ranked 8th out of 91 for burglaries.[1] *Burglaries too?!* The City of Pittsburgh Police cautions against comparing neighborhoods based on crime statistics, "Simplistic comparisons based only upon crimes that occur in an area do not take into account….factors that lead to a particular crime."[2] I get the caveat, I understand statistics, map making, and urban planning; but this is getting close to home. It's getting personal and uncomfortable. Is this my new normal: unsavory facts, gunfire, burglaries? I felt very far away from the affluent suburb where I grew up, where for 25 years we never locked the doors, and left car keys in the ignition.

That evening, I showed Joe the statistics on our charming, "Norman Rockwell" neighborhood. We talked it over again. At the end of it, we resolved to make our home a difficult target. I swallowed hard. I'd never given much thought to home security or self defense. Outside of our dogs, I didn't have a clue how to prevent a break-in, or defend my life if I had to. This suburban Queen of Denial checked herself into reality rehab.

My detached "It's not going to happen to me" script was supplanted with "Bad things happen to good people, Suzy. Deal with it!!" At 58, I'm still a tomboy and a Girl Scout at heart. I'm pretty strong inside and out. I've done a lot in my life and enjoy challenging projects, but this undertaking—to admit vulnerability, to revise my scripts, and adjust course—is the hardest thing I've ever done.

And so I set out to learn about the 'real world' and how to live in it with greater awareness and safety. In the process, I discovered the disciplines of T'ai chi and yoga, I enjoy expanded awareness and grace in movement; I breathe more deeply. I've learned to listen to and act upon my intuition, and unfailingly have better outcomes for it. In learning self defense and firearms training, I've discovered an ability for a steady, intense focus. Targets are hit. Self confidence elevates with practice. And at night, our doors are now always locked. For all of this, I live with less subconscious fear and finally sleep well at night.

I set out to learn about the world around me from highly qualified people in the fields of military, law enforcement, fitness, self defense, psychology, and philosophy. Their guidance has been kind and firm; their messages are clear and compelling. At the outset, I didn't know what to expect, but I can tell you this: my growth into proactive awareness has been unexpectedly liberating.

It's my intent to share this advice and guidance with women across the country. Using forethought and planning, we can prevent many crimes against our properties, our bodies, and our lives. Inside these pages you'll find real and actionable steps You can take to spare your property, save your life, protect your kids, and defend the elderly and disabled in your home.

> "A man's house is his castle and fortress, and each man's home is his safest refuge."

—Sir Edward Coke
The Institutes of the Laws of England, 1628

It's still not okay for someone to enter your home uninvited, thanks to a law penned in 1628 by Sir Edward Coke, an eminent English jurist and politician. Because of him, we have the metaphor, 'Your home is your castle'. This common law principle, established in England, was extended to the American Colonies and Canada. It stands to this day that no one may enter a person's home without permission.

—*Calamity May*

PART I

AWARENESS

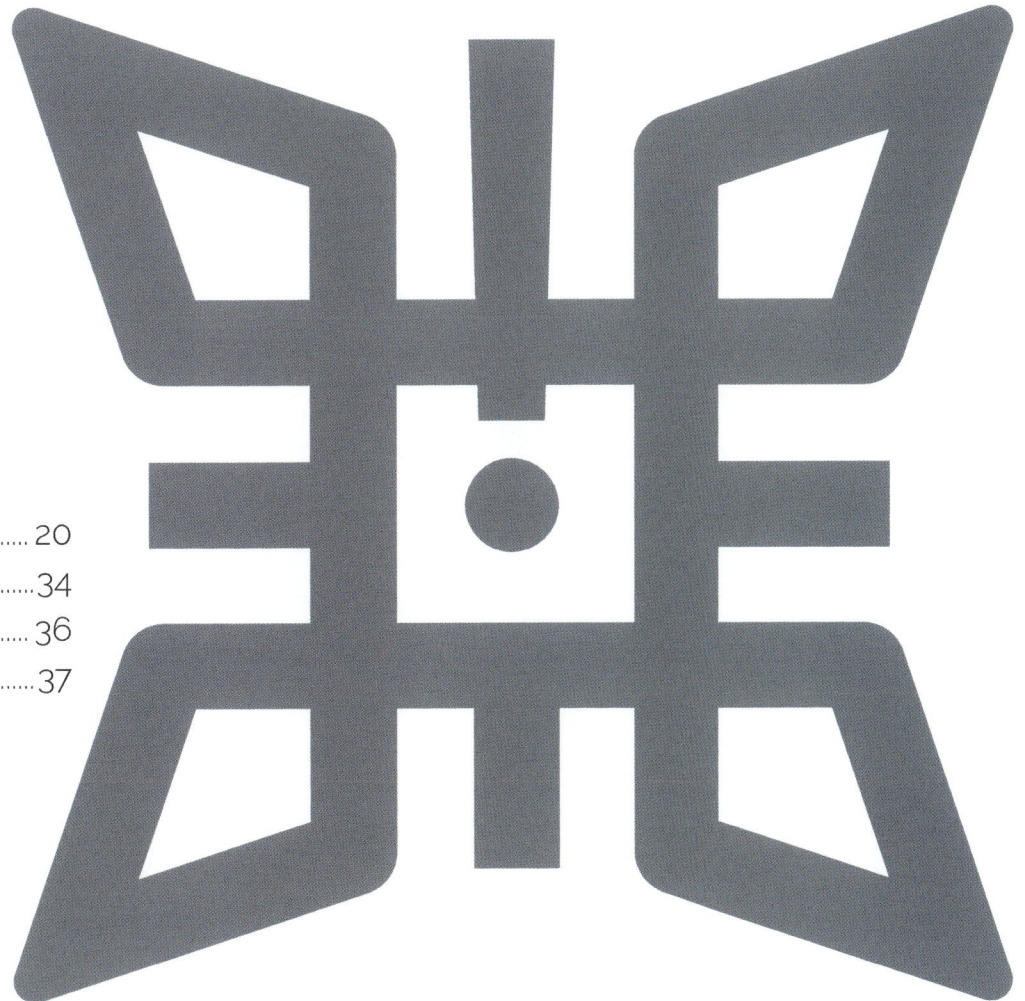

TABOO TOPICS

FAMILY & FRIENDS

I've unearthed a vein of fascinating, taboo topics among women and men—conversations we don't hear on TV or radio, over lunch or out to dinner. It started when I began asking friends and family, "Do you ever think about self-defense and protecting your home?" The answers varied from foot-shuffling silence, to a seam-busting moment when a woman explained her at-the-ready techniques in detail. Most of the time though, I find that women especially, can't, won't, or don't talk about it.

The conversations we're not having include: scenarios of "bad stuff" that we think could happen in our homes; what our reactions would be; tricks conjured up to deter an assailant; guns that some of us own; and, fear. Fear of what could happen, especially at night. That's the quietest topic, fear, as if to speak of 'it' might make 'it' happen. Hence, the silence.

I was reluctant to mention the undertaking of this book—what would my family and friends say? You're a landscape architect doing what?! What do you know about home safety and self defense? Did something bad happen to you? Luckily so far, I've not encountered perilous, bodily harm. Though in my 58 years, there have been small bumps in my path: the pervert in the woods when I was 9; Peeping Toms at my bedroom window when I was a teen; the naked guy in a raincoat making a frenzied gesture near my parked car; and a jealous boyfriend who shoved me against my kitchen wall, then later drove his car through my garage door. Restraining order obtained. (See Myth 2).

We all have stories about troubles encountered and troubles evaded. Almost every woman I talk to has at least one tip on how to avoid trouble, whether it's a 'walking tall' attitude learned from Mom, an intruder detection trick that Grandma used, or keys-as-weapons in your hand learned from a friend. People have been beat up, robbed, raped or worse for millennia. Women are spared none of it. Many of us have picked up tips and well-meaning advice from friends, siblings and significant others to spare us from such harms. Some are even useful.

Many possibilities exist for bad things to happen to girls and women, boys and men. In the mix of pre-teens, high school, first dates, college campuses, workplaces, parties, cars, hotel rooms, or the sanctuary of one's home, most women in their lifetime will be touched by negative experiences, violence, and/ or extreme physical danger. When I bring up these topics, I often sense a quiet reluctance or an emotional shudder.

In my work as a water resource planner, I have worked on projects that are designed to 'daylight' streams. This is an environmental engineering practice that identifies clean streams or springs that have been routed into sewer pipes, and restores them into the natural environment. Clean waters rescued from sewers flow again in sunshine and fresh air. With this in mind, it's my intent to daylight these taboo topics and the fears attached to them and bring them into the light.

Let's have conversations about being proactive, intuitive, and aware. Let's talk about products we can install to strengthen our homes; about appropriate choices for self defense; about defending ourselves and/or families in a dark, scary-as-hell moment.

Getting a handle on the real world, for me, started with recognizing my assumptions and the stories I tell myself. Let's face it, we all tell ourselves stories that allow us to adapt to, or rationalize the environments we live in, and lifestyles we've adopted. Left unexamined, these can lead to living life in an unrealistic bubble. Letting in real world facts illuminates how things really are. Trust me, it's not rosy switching from personal fictions to facts, but the pretense is over. The next 20 pages briefly establish real world context by using solid numbers and references to American law. Then I bring it home...

VULNERABILITY

Since 1880, the United States has conducted a census of its citizens every 10 years. The last one was in 2010. The census counts heads, tallies males and females, asks how many kids are in a family, how many people live in each house, their age and relationships therein. In broad strokes, it captures America's domestic arrangements and ethnic composition. Census 'snapshots' over 140 years yield a robust collection of data used by public agencies and private companies to determine trends. One long uptrend is women as "householder," defined by the Census Bureau as the person "in whose name the home or apartment is owned, being bought, or rented."

According to the 2010 census, many women, of all ages, live alone in the U.S. One in three homes, (36 million residences) are paid for by women, with or without kids; single or divorced; young and old. The greatest uptick is 'empty nesters', American females born during the post-World War II baby boom (peak year 1957) who grew up, got married, had families, then their grown kids moved out. Between the 2000 and 2010 censuses, the number of empty nesters jumped 29%.

Yes, living solo can be sweet, and you've done a lot already...isn't it time to rethink your own personal safety?

Why? Because being the head of a household means keeping the roof over your head, bills paid, food in the fridge, ear to the ground, and assuming responsibility for, (or negligence of) the safety of your home and lives within it. Crimes committed against people and property stack up against those who live alone in urban areas. Those most prone to crime are poor, female-headed households in large cities.[1]

AMERICAN HOUSEHOLDERS

American households in the U.S. in 2010:	116.7 million		
All female heads of households (with or without children) who live alone, or with a roomate or domestic partner	36 million	31%	
Women who live alone, no children:	17 million	15%	
Women with children and no spouse:	15 million	13%	
Women whose kids have moved away and who now live alone:	7 million	6%	
Women and men, 65 years or older who live alone:	11 million	9%	

EMPTY NESTERS UP 29% FROM 2000

U.S. Census Bureau, 2010

Regardless of money or relationship status, most Americans are deeply uninformed about the extent of crime in the United States. As a result, most of us are underprepared to prevent a home break-in and possible harm. Before revealing the Department of Justice's numbers on crime, I'd like to point out a problem with crime statistics: the numbers we see and hear about don't reflect reality. There's far more crime than what's reported. The vulnerability factor just went up!

UNDER-REPORTED CRIME

It's hard to gauge how dangerous things are because nobody wants to reveal or talk about real crime numbers. Under-reporting crime isn't new, it's long been practiced. Under-reporting of crime is undertaken by citizens, law enforcement, politicians, and marketing consultants to name a few. Mainstream crime statistics that we see about cities, states, and the nation are, to varying degrees, artificial constructs used to create impressions of safety and livability; bolster citizens' confidence; and, attract economic development.

> Crime statistics aren't what they appear to be.

It starts when a person experiences a crime or attempted crime (against property or person) and doesn't report it to police. It's the duty of sworn officers to respond to calls about active or recent criminal activity; talk with victims to investigate details; take a report and turn it in. Police then work to apprehend the person(s) responsible for the crime(s). Responding to crime is the most significant role for law enforcement regarding public safety. However, officers cannot respond to, or file reports on, crimes not brought to their attention. Reporting a crime to a co-worker, head nurse, human resources, or clergy, for instance, won't 'count' as a crime or attempted crime until it's reported to police. As members of society, if each of us doesn't report crimes, threats, or thefts, we'll never know the extent of crimes in any community, city, or campus.

There are 2 broad categories of crime: 'crimes against people' and 'crimes against property'. Let's look at several groups that have very different motives for not reporting crimes against people.

VICTIMS Of the victims of violent crimes—women and men who experience rape, or attempted rape; who were beaten, threatened or injured with a gun, knife, or dangerous 'personal weapon' (hands or feet)—it's estimated that 50-88% do not report the crimes against them.[2] Four top the list: rape, domestic violence, hate crimes, and sexual abuse of children (under 12 years old). Rape and attempted rape are considered to be the most underreported crimes of all.

The biggest reason for non-reporting is well known: most victims know their assailant and fear reprisal or worse treatment if the crime is reported. Other reasons: a victim feels shame; suspects retaliation; doesn't want the assailant to get in trouble; or, feels apprehensive about potential treatment by police officers.[3] It's almost understandable, but each unreported crime is unfortunate for our neighborhoods and society. Victims go unserved. Criminals get away, sometimes with horrendous acts. The most reliably reported violent crime is homicide—it's hard to ignore a dead body.

UNREPORTED CRIMES IN 2014[4]

Unreported violent crimes against people: .... 5.4 million

Rapes: 284,000

Robberies (taking property with threat of weapon or inducing fear): 664,000

Aggravated assaults (threat with a deadly weapon): .... 1.1 million

Simple assault (threat without a weapon): .... 3.3 million

Burglaries: .... 3 million

Thefts: .... 11.7 million

LAW ENFORCEMENT AND POLITICIANS   At the last tally in the United States, 17,985 law enforcement agencies were counted at federal, state, county, city and municipal levels; public colleges and university campuses; and, tribal lands.[5]

Underreporting within law enforcement can happen like this: Politicians and administrators who want their city, or campus, toappear safer than it really is, send 'word' to law enforcement's upper hierarchy to reduce crime statistics. In turn, on-duty police deal with crime numbers accordingly. For instance, when a victim reports a crime, a sworn officer shows up, investigates (or apprehends a criminal in the act), and writes up a report. If that officer is faced with pressures 'from above' or has a not-to-exceed quota, he or she may downgrade a violent crime (felony) to a lesser crime (misdemeanor), or fail to file the report at all (shredder). The TV series, *The Wire*, exemplifies this practice, in Season 3, Episode 1, "Time after Time".[6]

Underreporting crime is data manipulation. It's "unethical and likely illegal"[7], and happens every day in some law enforcement entities across the United States. But there's one more step: crime reports received by law enforcement agencies, are voluntarily submitted to the Federal Bureau of Investigation (FBI) where crimes are summarized into the FBI's Uniform Crime Report (UCR) system. In turn, annual crime reports are published.

Eight types of crimes comprise the backbone of the UCR. The FBI considers these offenses to be "serious crimes [that] occur with regularity in all areas of the country." They are: Criminal Homicide, Forcible Rape, Robbery, Aggravated Assault, Burglary, Larceny-theft, Motor Vehicle Theft, and, Arson. (See definitions in Appendix A). The FBI's Uniform Crime Report is widely used and extensively cited by policymakers, politicians, police associations, academics, and media to understand regional and national crime trends. (Remember, only reported crimes make it into the UCR).

The next time you read about "a significant drop in violent crime" or "consistently lower crime rate since Mayor X took office three years ago", the claim may be legitimate, or it may be falsely propped up. It's hard to know given that the FBI audits only 1% of those 18,000 law enforcement entities each year.[8] Internal audits are laudable, but rare. Underreporting tactics have long been with us, and they continue to be popular with new mayors, their hand-picked police chiefs; and, university presidents, to name a few.

Why do law enforcement agencies do this? A co-mingling of words come to my mind, broad themes that apply to communities, large and small: Politics;

TABLE 1. UCR CRIMES RELATIVE TO MY CITY[9]

City	Population	Rapes Jan-Jun 2012	Rapes Jan-Jun 2013	Burglaries Jan-Jun 2012	Burglaries Jan-Jun 2013
Anchorage, Al.	299,000	145	199	503	591
Anaheim, CA	345,000	47	39	725	784
Bakersfield, CA	356,000	21	27	2,237	2,191
Riverside, CA	314,000	36	35	1,060	1,075
Santa Ana, CA	332,000	25	31	503	452
Aurora, CO	334,000	103	117	920	952
Tampa, FL	351,000	18	40	1,115	920
Lexington, KY	302,000	60	61	1,343	1,226
St. Louis, MO	319,000	101	162	2,262	1,960
Greensboro, NC	276,000	34	42	1,737	1,344
Newark, NJ	279,000	26	23	1,019	1,048
Henderson, NV	263,000	35	16	629	669
Buffalo, NY	262,000	75	73	1,766	1,621
Cincinnati, OH	296,000	90	83	2,734	2,448
Pittsburgh, PA	**312,000**	**19**	**30**	**1,189**	**1,072**
Corpus Christi, TX	313,000	87	72	1,170	1,284

City populations are rounded up or down, and range from 260,000 to 360,000. Cities are arranged alphabetically by state.

Rape: "slightest penetration" against a woman's will.

Burglary: unlawful entry of a structure to commit a felony or a theft, via force, no force, or attempted forcible entry.

FROM THE HEADLINES....

Chicago – Chicago Magazine, June 2014
The Truth About Chicago's Crime Rates, Part 2
"You don't need to be a rocket scientist to realize that not counting nearly 60,000 [UCR] index crimes can do wonders for a city's overall crime rate.... from 152,031 in 2010...to 86,174 in 2011".
[In 2011, Chicago saw Rohm Emanuel sworn in as the new mayor and Garry McCarthy, the new Superintendent of Chicago Police.]

Los Angeles - LA Times August 10, 2014
LAPD Misclassifies Nearly 1,200 Violent Crimes as Minor Offenses

Yale University - Business Insider, May 16, 2013
Yale Fined $165,000 For Failing To Report Sex Crimes

31 Large Universities and Colleges - Huffington Post, February 3, 2015
Study Shows When The Feds Leave, Colleges Go Back To Underreporting Sex Assaults

See Appendix B for more headlines

Marketing (for economic development tourism, college admissions; real estate values), and Managing Perceptions of Safety. With 'reasonable' numbers on the books, mayors are happy, police chiefs keep their jobs, citizens read about reduced crime rates, and marketing consultants tell the world about a city's livability. (New York City's former Mayor Bloomberg had a favorite saying, "the safest big city in America", a statement known to be propped up with 'adjusted' crime numbers).[10] A few newspapers headlines, above, point to underreporting. (See more in Appendix B).

If you want to get a handle on crimes on your campus or where you live, it's most likely that you will not get the full picture. I have yet to see how to assess the degree of 'truth'.

I wanted to get a feel for the UCR data, so I did a simple crime comparison of small cities with populations similar to my hometown of Pittsburgh, PA. Online, I looked up the FBI's 2013 "Preliminary Semiannual Uniform Code Report" and selected cities with comparable populations (plus or minus 50,000), and that report UCR crimes to the FBI. I focused on burglaries and rapes over the same 6-month periods in 2012 and 2013. It took me about an hour and a half to compile this list of 16 cities. (See page 15).

If I move from here, I would consult the most recent UCR for an impression of safety in cities, and double the crime numbers. (Mind you, this is me talking). Why the doubling? Two reasons: gross underreporting by victims (58%

underreporting across all crimes)[11], and downward manipulation of crime statistics by law enforcement. In large cities of several million people doubling published crime numbers may be conservative.

Unfortunately, this story's not over. Though it's been in the news, and is increasingly documented and reported, rape, attempted rape and other sexual assaults are rampant in two of America's hallowed and trusted institutions: university campuses and the American Armed Forces.

(If you ask, "How is this relevant to home security and self defense?" I have this answer: Despite their high levels of public trust, many schools and military branches tolerate an underreported body of rapists and serial rapists, and turn out many victims of sexual violence. Sexual predators circulate in our society. It's information and context).

COLLEGE CAMPUSES On May 1, 2014 the Department of Education released, for the first time ever, a comprehensive list of 55 colleges and universities that were, at the time, under review for possible Title IX violations—specifically, mishandling rape, attempted rape, and sexual harassment incidents on campuses. In January 2017, that list grew to 225 colleges and universities. (Appendix C shows you how to request the most recent list). Learning institutions, from high schools to private elite universties, that receive federal funds must, under the Jeanne Clery Act, report crimes and attempted crimes—sexual violence, dating violence, domestic violence, and stalking—committed on campuses and in on-campus housing.

Sending teenagers like Katy or Jackson off to college is akin to a rite of passage: they leave home, grow up, learn responsibility, take in new information, and ideally become informed, skilled and employable citizens. Nobody sends their kid to school to take on sexual predators! Rape, Abuse, and Incest National Network (RAINN) and the Department of Justice both estimate that in 2014, 80% of all female victims attending schools of higher education do not report

HIGHER EDUCATION CAMPUS CRIME, PENNSYLVANIA[12]

Reported crimes on campuses and on-campus student housing, for Pennsylvania schools of higher education that receive federal funding:

	2004	2012	2013
Forcible Rape	283	522	761
Burglary	1962	1331	1155

**It is estimated that 80 to 88% of campus-related rapes go unreported.

In a public statement, Prof. Caroline Heldman, Ph.D. called the underreporting of rapes and attemted rapes on American college campuses a problem of "epidemic proportions".

—June 14, 2014
Boise State
University

rape or attempted sexual crimes against them to police.[13,14] Rape victims who are students are less likely to report than non-students. This is a great misfortune for four reasons:

First, the victim does not avail herself (or himself) of really helpful services to deal with the trauma. Second, unreported sexual predators / offenders continue to circulate through school or teach classes and may never be called out. Third, sexual assault numbers this large call for major re-structuring of rape prevention campaigns. Finally, we all lose here, because unidentified serial rapists and their current addresses will never be made known via online public sex offenders lists, like Megan's Law, or on Family Watchdog, both display maps of registered sex offenders near you.

This alarming story of sexual predation on college campuses is amplified, even more, in all branches of the American Armed Forces and military academies.

AMERICAN MILITARY A friend told me about a documentary called, *The Invisible War*, hosted at her Unitarian Church, about rape in the military. I went and took a seat next to Gwen, a friend and psychiatrist who specializes in depression. We talked about her daughter's first term up at Penn State, our own alma mater. The film started. An hour later the lights came on, we looked at each other. She nodded her head knowingly. I'm pretty sure my jaw dropped all the way open.... I had no idea things like this happened in my country.

In 2012, across all of America's Armed Forces, the Department of Defense estimates that 26,000 sexual assaults occurred but only 3,374 were reported. Their 2012 estimates of sexual assault were up by 35% from 2011.[15] The numbers of men raping men are proportionately higher than men raping women, because there are so many more males. There's even an acronym to describe the psychological effects on victims: MST for Military Sexual Trauma. To begin to understand how this can happen to men and women who choose the military as a career path, or how the Armed Forces can get away with these terrible acts, I highly recommend watching *The Invisible War*.

Young cadets, men and women, in military training academies (Air Force, Military, and Naval) are also vulnerable. For the year ending May 15, 2015, 91 reports of sexual assault and harassment were filed, up 55% from 2014.[16] Victims are speaking up.

Arizona Senator John McCain, himself a Vietnam veteran, pilot, and ex-prisoner of war is unable to recommend the military as a career path or call to duty, he told the uniformed chiefs of the Army, Navy, Air Force, Marine Corps and Coast Guard at a Senate Armed Services Committee hearing:

"Just last night, a woman came to me and said her daughter wanted to join the military and could I give my unqualified support for her doing so. I could not... I cannot overstate

SEXUAL ASSAULT AMONG MILITARY PERSONNEL[15]

Pentagon's estimate of "unwanted sexual contact", 2012: 26,000

Number of reported sexual assaults (rape and attempted rape), 2012: 3,374

Pentagon's estimate of "unwanted sexual contact", 2011: 19,000

% of Unwanted sexual contact among Armed Forces servicewoman, 2012: 6.1

Includes active duty members Army, Navy, Marines, and Air Force

SEXUAL ASSAULT IN THE SERVICE ACADEMIES[16]

Unwanted sexual contact among female cadets & midshipmen in

US Air Force Academy, 2012: 11.2%

US Naval Academy, 2012: 15.1%

my disgust and disappointment over continued reports of sexual misconduct in our military".[17]

As of December 31, 2013, there were 1,370,000 enlisted people on active duty across all U.S. Armed Forces. These soldiers stay in the military on a career path, or leave to take jobs in many different civilian sectors, like law enforcement and private security. Some are perpetrators; many are victims coming out of a culture of rape, harassment and intimidation. The ongoing campaign of domination and violence in these American institutions is a paradigm in need of nothing short of radical reform.

It's not a stretch to say that there is more real crime than we'll ever know (or may want to know). And it's likely to go unchanged, for now. So, reader, learn to be responsible for your personal safety. Harness your intuition. Cultivate your survival instincts. Lock your doors. And learn real world self defense.

PROPERTY CRIME Let's go to the places where we live and look at 'crime against property'. In this book, we consider burglary—the unlawful entry of a residence, or outbuilding, with the intent to steal, or commit a felony (a violent encounter). Police describe how a bad guy gets into a home using these terms:

Google makes it easy. A search for "burglars walk in front door", yields these searches:

- how burglars pick a house
- how to break into a locked window
- how to break a house window silently
- best way to break into a house
- most common burglary entry points

1)   **Forcible entry** (lock tampering; kicking in a door)

2)   **Unlawful entry**, no force is used (entry through an open window, or unlocked door)

3)   **Attempted forcible entry**

What's the difference between theft and robbery?

Theft is burglary without force or violence, (a lawnmower stolen from the garage; jewelry from the dresser; a plasma TV from the wall; pharmaceuticals from the bathroom).

Robbery uses force, violence, or threatening behavior to steal something of value from a person (a wallet taken at gunpoint; or being threatened in your home to produce the key to a lockbox).

In 2014, 1.73 million burglaries were reported nationally.[18] Official estimates of unreported burglaries nearly double that number.[19] Common reasons for not reporting burglaries, include: homeowners find it a nuisance to report a crime; they think the stolen goods aren't worth the trouble; they don't want police at their house; the burglar is a family member. Or, all of the above.

Bottom line: there's significantly more real crime in the US than we'll hear about from newspapers, radio, or TV news. And it's unlikely to change.

When I see the real numbers of crimes against people and property, and consider the ranks of vulnerable people–women who live alone, young people away at college, and men and women serving in the Armed Forces, I feel like I'm standing at the intersection of Unsuspecting Avenue and Predator Way.

If you're as shocked or dismayed as I was, stick with me—there's more solid ground to cover. What a compass is to magnetic North, facts are to the real world. They got me out of Denial Territory and on the road to Awareness. Reading the five following misperceptions may help you adjust course. They're commonly held beliefs that many of us wrap ourselves in like warm blankets on a cold night—except they're not real, they're myths.

FIVE MYTHS

Dangerous assumptions you need to rethink.

It's what you learn after you know it all that counts.

– John Wooden, basketball coach

MYTH 1

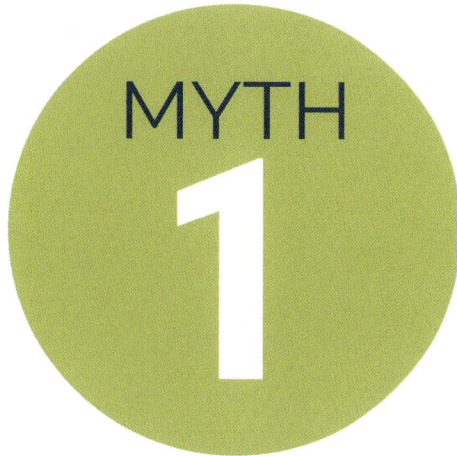

The police will protect me.

REALITY

Police have no legal obligation to protect you. They are not authorized to protect you or your family from violence; nor protect you from threats or harassment; nor assist in the protection of your property. Police are not mandated to protect citizens from the bad actions of other citizens, or prevent them from occurring. Furthermore, police departments and their officers cannot be sued for failure to protect.

"He's going to kill me....You [police] never do anything about him. You talk to him and leave."

—Nicole Brown Simpson
L.A. Times, June 17, 1994

BOTTOM LINE

Every American—woman, man, and child regardless of age or ability—is responsible for her/his own personal safety.

AMERICAN LAW

United States Supreme Court

Chief Justice Wm. Rehnquist pens: "Nothing in the language of the Due Process Clause itself requires the State to protect the life, liberty, and property of its citizens against invasion by private actors [one citizen's bad or harmful actions against another-SM]. DeShaney v. Winnebago Dept. of Social Services. Citation. 489 U.S. 189 (1989).

State Laws

"The government is not liable even for a grossly negligent failure to protect a victim of a crime". Riss v. New York, 22 N.Y.2d 579, 293

"Neither a public entity [police-SM] nor a public employee [policeman/woman-SM] is liable for failure to....provide police protection service or, if police protection service is provided, for failure to provide sufficient police protection service". California Government Code Section 845.

See Appendix D for more examples and interpretations.

MYTH 2

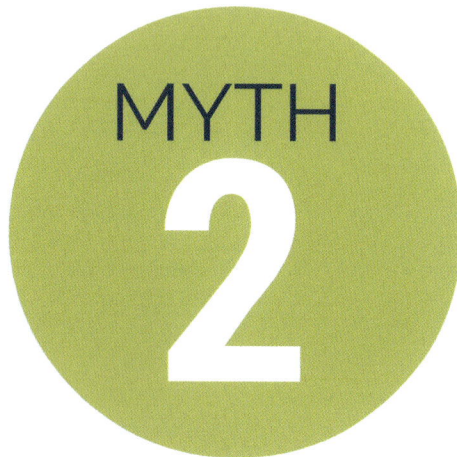

I have the protection of a restraining order.

REALITY

Court-ordered restraining orders do not mandate police protection. Victims of domestic violence, or victims of threats, harassment or stalking, who seek restraining orders to protect them from harm have no constitutional right to police protection, nor enforcement of a restraining order. Furthermore, police cannot be sued for lack of protection no matter the degree of violence, or even death. This applies to children named on a restraining order.

"A restraining order isn't worth the paper it's written on".

— Jessica Lanahan, formerly Jessica Gonzales

BOTTOM LINE

You are responsible for your own personal safety, and the safety of your children.

AMERICAN LAW

United States Supreme Court

The landmark case of the Town of Castle Rock v. Gonzales has far-reaching impacts for women, men, and children who have been subjected to domestic violence and seek protection from abuse through restraining orders. In 2005, The United States Supreme Court (545 U.S. 748) decided in a 7-2 vote that Jessica Gonzales, the mother of three slain children had no constitutional right to police protection or enforcement of a restraining order, even though she lived in a state that required mandatory arrest for a violation.

Read this tragic, riveting story and the verdict reached by an international tribunal in Appendix E.

MYTH 3

The police will arrive in time to protect me.

REALITY

Most illegal entries, burglaries and/or violent incidents in the home happen quickly, within 5-10 minutes.[20] By the time you call 9-1-1, or your alarm monitoring company calls them, many factors (day of the week, time of day, proximity to police station, other calls in progress), determine how long it takes for police to respond. Serious crimes in progress get priority. In most cases, criminals flee before police arrive.

"It gets ugly if somebody's home."

—Brian Keith, retired law enforcement officer

BOTTOM LINE

If there's time to call 9-1-1, and you are home waiting for police to arrive, be prepared to save your life. Have a plan!

For the vast majority of crime calls, the only value to a fast response is if it's really fast. And for about 90% of crime calls made to police, it's literally impossible to respond this quickly.

—William Spelman, Professor, Lyndon Johnson School of Public Affairs, University of Texas, Austin

FACTS

There is no national standard for measuring response times to 9-1-1 calls. Police have discretion about if, and when, they respond to 9-1-1 calls that are not crimes in progress. Factors that can affect how long you wait:

- Ratio of on-duty sworn officers to citizens.
- Population density. Residents in dense cities have faster response times than those living in rural areas.
- Proximity to a police station or sheriff's office.
- How occupied dispatchers and police are when your call's received.
- Seriousness of the crime called in. Serious crimes in progress get priority.

Example of police response times in relation to priority:[21, 22, 23]

	Priority Crime	Low Priority Crime
Milwaukee	14 min	
New York City	7 min	17-18 min
New Orleans	20 min	
Denver	11 min	20 min
Detroit		60+ min
Dallas	8 min	22 min

MYTH 4

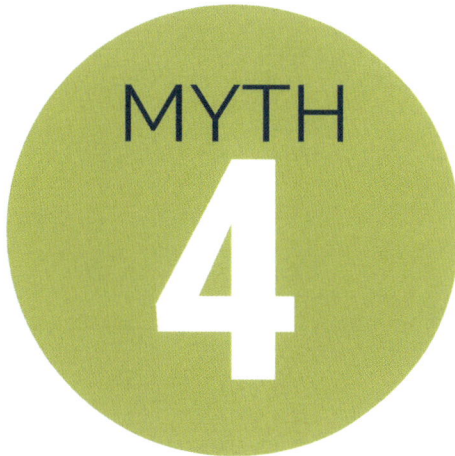

My home security system protects me.

REALITY

Home security systems do not bar intrusion; they offer no protection. They can alert you to the presence of a real or potential break-in. Today, many cities, require verification (from the homeowner) to determine if it's a real home break-in and not a kid or dog setting it off. If it's a real home intrusion, the security company will call 9-1-1. If you're home during a break-in, home security companies cannot and will not advise you how to respond to an intruder.

"I want to know if an intruder is on my property before they get to my doors and windows."

-Calamity May

BOTTOM LINE

If you are home and alerted to a breach, you may have long minutes to wait time until police arrive. This is a crucial time. Have a Plan and Backup Plans.

FACTS

If you pay for a monitored home security system, turn it on, and have a real home invasion, expect a chain of notification to take place and wait for police to arrive.

Many police departments will not send policemen to a residential alarm unless the homeowner is reached and verifies illegal entry. Police departments in Salt Lake City, Milwaukee, Dallas, and Las Vegas, for instance, only respond to verified calls.

Nationwide, most home security systems register an average false alarm rate of 95%. Police departments in an increasing number of American cities don't respond to home alarms because their resources are spread too thin. Examples include: Los Angeles, Detroit, San Jose, Savannah, and Seattle.

POSSIBLE

MYTH

5

My husband/partner will protect me.

REALITY

If you're fortunate to have a capable partner who's home when you are, is physically and mentally capable, sober and/or un-medicated, remembers where stuff is, knows how his/her defense tools work, keeps them maintained and at the ready, and doesn't sleep like the dead, and who's willing to be a dedicated home defender, then No, that's not a myth.

"Maybe you have a knight in shining armor... If not, what are **you** willing to do?"

—Calamity May

BOTTOM LINE
While of sound body or mind, never relinquish your personal safety to someone else.

FACTS

Countless Americans may be unaware of, or, incapable of responding to a home break-in:

- 90 million Americans suffer from snoring activity during sleep.[24]
- 32 million Americans (nearly 1 in 7 adults), struggled with a serious alcohol problem in 2015.[25]
- 16 million Americans report using white noise machines in their bedrooms.[26]
- 9 million Americans use prescription sleeping pills.[27]
- 1 in 8 Americans over the age of 60 reports worsening memory loss.[28]
- How many sleep with window air conditioners on and doors closed?

BIG TAKE AWAY: You may need to be your own first responder.

GOOD PRACTICES

AWARENESS

Humans are part of the animal kingdom. What are wild animals constantly doing? Watching each other, cognizant of other animals and potential threats. They graze with one eye on the horizon. For at least 400,000 years, Homo sapiens have been watching each other. Family, friends, and predators—we watch each other all the time. We're still vested in figuring out who's weak and who's strong; who's desirable as a mate and who you need to be concerned about.

Animals also remind us (excluding the very young) that each of us is responsible for her/himself. An individual animal doesn't make assumptions that its mates, offspring, or friends are going to protect or defend it. So too are you responsible for your safety and well being. Your personal safety is yours and yours only. While you're well in body and mind, your safety is not to be abdicated, ignored, put off, or given to someone else to look after.

In the course of this book, I've met former Marines, Rangers, and Navy Seals, firearms and martial arts instructors, individuals with 4+ black belts, men once robbed by strangers, and women raped by men they dated. They ALL begin the conversation about preventing harm and avoiding violent behavior with Awareness. To be Aware, you have to pay attention, to varying degrees, much of the time.

Every day we are immersed in sights, sounds, smells, touch, and temperatures. Our brains process information and make conscious and unconscious decisions accordingly. Too cold at work? Put on a sweater. Something burning in the kitchen? Go investigate. Too hot in the sun? Seek shade. Thirsty? Drink water. Nothing new here. We act upon these things more than we think about them.

Life decisions are a lot more complicated. Every day we make choices regarding that day and into the future about ourselves, our families, homes, work, food, income, health, love, or retirement, or the lack thereof. That's a lot of thinking.

On top of that, most of us distract ourselves from all that thinking by immersing ourselves into personal technologies and communication: Internet, phone, texts, and tweets; 24-hour news and sports; celebrity dramas; global wars; ads for junk food and weight loss; Facebook likes and LinkedIn updates. We walk with earbuds in, and text while driving. Let's face it, most of us embrace these personal technologies as 'inevitable' necessities. Ironically, many of us seek out distractions (like shopping) and addictions (like alcohol) to buffer ourselves from the above Immersion Channel. The problem: this constant information barrage compromises one's situational awareness in the real world. Many people are hyper-connected and not paying attention at the same time.

What's going on around you right now? What events and movements are taking place? Who's in close proximity to you? What's happening in your peripheral vision?

You can go a long way toward preventing bad things from happening to you by being present, mindful, and clear headed. Be aware of the physical, social, and the energetic contexts of places you visit. Whether waiting at the bus stop, staying late at work, or standing in line, part of you needs to be practicing awareness, discerning input, and making good decisions.

A book could be written on avoiding trouble. Why this is relevant is because where you go, who you hang with, how you act, and how you carry yourself impacts your personal safety.

While writing this book, I've made it a point to practice awareness in different ways each day, and find small, rich rewards in the process: I gave up multi-tasking (was never good at it) and uni-task instead with better results. When I walk, I immerse myself in the present—the landscape, sky, plants, and people (instead of my problem or daydream du jour.) When I talk with someone, I pay attention to what they're saying (instead of hearing with a hurried mind or jockeying for position). I get more from conversations and the people I'm with.

My personal awareness training has expanded to where I take in significantly more details about places, environments, and peoples' faces, eyes, energies and demeanors. I perceive positive energies and register dark ones. I find it remarkable to take in so many more details. This input doesn't make me paranoid, in fact, I let this information drop without alarm. As I gain proficiency and attain greater capacity for more awareness and input, I exercise a great gift that each and every one of us was born with, intuition.

> Don't do stupid things,
> Don't go to stupid places,
> Don't hang out with stupid people.
> – John S. Farnam

INTUITION

Noreen Renier is a successful psychic detective who's helped police solve hundreds of missing person cases and cold case homicides. She writes in her book *The Practical Psychic*, "your intuitive mind is constantly working for you, although you may not know it." She speaks to that which diverse cultures have known, and famous people have embraced about this exquisite and stunningly accurate aspect of the non-rational mind: "Your mind is a wonderful tool, a resource you can apply in any way you wish. You can opt to use as much or as little of it as you want. You can limit or expand your consciousness….[and] develop abilities using intuition and logic to use more of your brainpower than most people do."

The Chinese word for intuition translates to "direct heart"; in Japanese it's called "haragei" meaning "belly talk"; in Hebrew the word Daat, means where inner knowledge / inner teacher resides; in the Lakota language a wise intuitive person is "he or she that goes beyond knowing." Carl Jung defined intuition as "perception via the unconscious." The Briggs-Myer personality test, an offshoot of Jungian archetypes, uses intuition as one of the dominant, determining traits in how some people perceive information and solve problems.

Logic and rigorous analysis drive the ways in which our brains process data and crunch numbers. When one makes important decisions however, it's intuition that can fill in gaps missing from spreadsheets and charts. When asked about how he makes the best decisions possible, Bill Gates, co-founder of Microsoft said "often you have to rely on intuition." Apple's co-founder, Steve Jobs, describes his discovery of intuition as a teenager in India, "The people in the Indian countryside don't use their intellect like we do, they use their intuition instead, and their intuition is far more developed than in the rest of the world. Intuition is a very powerful thing, more powerful than intellect, in my opinion." Another remarkable person was Buckminster Fuller, a prolific inventor and visionary famous for synthesizing ideas across many disciplines of science, math, and engineering. He named his sailboat *Intuition*.

Each of us has this gift of intuition. On a daily basis it conveys a 'sense' of things that we don't directly perceive or know firsthand. For instance, you 'sense' being stared at from a distance; 'feel' someone standing behind you; 'know' who's calling when the phone rings. Some refer to intuition as a gut feeling, a personal radar, or 'the little voice inside'. Whatever you call it, it's a finely tuned instrument that instantly detects the presence of others and alerts you to malicious intent and/or impending danger. Equally, it can discern safe environments and positive people. We're hard wired for intuition, it's part of how we're made and the switch is always On. It is the brain's job (logic) to respond to the blips that show up on this internal radar and make decisions or take action. Intuition can also trigger the body to physically move out of immediate danger.

Sense what's happening before it happens, and get out of the way.
— My Tai Chi Teacher

Trusting your intuition is the exact opposite of living in fear.
— Gavin de Becker

Gavin deBecker, heads a security consulting firm whose clients include presidents, prime ministers, and Hollywood elite. He writes the insightful book *The Gift of Fear*. In it, he articulates many aspects of intuition as an invaluable ally to be trusted explicitly, calling it a "tool for personal safety and awareness." He differentiates between survival signals conveyed by intuition from stuff we make up: manufactured fears, worries, and anxieties. He teaches us to identify, respect, welcome, and trust intuition and its wide range of messages. He implores us to make room in our lives for intuition and cultivate its power to live a safer life.

He points to one downside of intuition: if you don't pay attention to it, it's of no use. Many Westerners deny its relevance, blow it off, and disregard its message. Worse, they dismiss its real and imminent survival signals. To this deBecker says, "No animal in the world suddenly overcome by fear, would spend any of its mental energy thinking, 'It's probably nothing'." If your gut tells you to Do this, Take that, Stop! or Go!....Do That Thing. If you find yourself on a city sidewalk with three guys walking toward you and your instinct is to cross the street, **act on it**! Develop trust in your intuition. Cultivate its brilliance. Be amazed how powerful an ally it is.

There's one more cornerstone to cover, it's powerful, free, and open source: how you physically carry yourself in the world.

DEMEANOR

The image you project and the body language you convey, reflects who you are whether you're conscious of it or not. Your presence and energy has huge impact on how, or if, you're approached by strangers, and how you're treated by people you know. Upon first encounters, we assess each other in split seconds. Salesmen size up customers; schoolteachers suss out students; predators assess potential victims looking for easy prey. How you walk down the street conveys vulnerability or strength, meekness or confidence. One look, and a stranger may accurately determine if you're a pushover or a fighter. A purposeful stride, shoulders back, and head up conveys confidence. Conversely, a shuffling walk, shoulders forward, head down and avoiding eye contact conveys meekness, as if you've already given up. Be aware the signals you send.

If you're wondering what this has to do about preventive measures in your home or self defense....I have two answers: the strengths or weaknesses you convey may flow over into where you live; and, predators follow some people home based on perceived vulnerabilities, then assess their homes or apartments.

Stand up straight and realize who you are. That you tower over your circumstances. You are a child of God. Stand up straight.

– Maya Angelou

Demeanor is a smart awareness exercise to cultivate everyday. In a nutshell: Stand up straight, head up, walk with purpose. Walk like you have an important destination and someone waiting for you. Look people in the eye (but don't stare). You'll notice a difference! Try it at home, try it at work, try it when you're out shopping. If it feels funny, well, *Fake it til you make it!* Look confident then become confident. You'll feel differently about yourself; people will notice that you have presence, strength, and backbone.

This brings us to the time-honored tips below. Self defense experts agree on these street savvy tips:

DO

Walk tall, shoulders back, spine straight.

Hold your head up.

Make eye contact.

Look self assured, confident.

Pay attention to your surroundings.

Respond to a stranger's request by speaking assertively. It's okay to appear rude if you feel uneasy about a situation or person.

Park your car close to your destination and under lights (especially at night)

Heed your intuition. Trust your gut.

Act on your instincts.

"No" means No. Say it like you mean it.

DON'T

Don't look at the ground

Don't avoid eye contact, but don't stare

Don't look wimpy, weak, drunk or stupid

Don't wear earbuds or text on your cell phone while walking or in transit.

Don't be compliant with someone that makes you feel uncomfortable.

Don't walk long distances across parking lots; don't park in dark places; avoid parking garages when possible.

Don't fumble for your keys as you approach your car, house, apartment, or work.

THINK LIKE A PREDATOR

RECOMMENDED: AMY CUDDY'S TED TALK

" Your Body Language Shapes Who You Are".

Amy Cuddy is a social psychologist who describes the power and science of "power posing". Before a big meeting, performance review, or interview, standing like Super Woman with your hands on your hips prior to that event, can affect testosterone and cortisol levels in your brain, and crank up your confidence level. Find it online.

Immersing yourself into how a criminal thinks can be instructive in helping you put together, or add to, your prevention and defense system.

Most of us are good people. Even as more and more Americans face increasingly difficult financial times, most don't go out and steal, rob, or prey upon others, only a very small percentage of the American population has a record of doing that. To gain insight on those who do, we can briefly access the 'dark side' through role playing. Two exercises below, allow you to practice how a new or seasoned criminal sizes up potential victims and narrows down choices to those who are vulnerable or incapable. It's surprising how quickly we take on these unaccustomed roles! Try with friends or family members. You may see things a little differently.

ROLE PLAYING: WALKING THE MALL

Go to the mall on a busy day and act like you're there to shop. Turn off your cell phone. Walk and merge with the pack, or sit on a bench if you need to, and observe. Focus on being a predator at work: sizing up sheep, looking for easy prey. Look at the people around you, not so much their clothing, hair, or shoes. Instead, note how individuals carry themselves. Some people appear strident, confident, self assured. Others shuffle along. What qualities do individuals convey? How do the following qualities affect your perception of a person:

- A person's posture, how does she/he hold their shoulders or head?
- What do you read into their stride?
- What kind of character do they exude?
- Do they project confidence, strength, and/or self esteem?
- Are they fit or out of shape?
- Are they distracted by their cell phone?

From the back of your predator brain, recall that height or build doesn't matter. A small person can be confident and capable, even mighty.

Identify the people in the herd who appear meek, weak, and vulnerable. Ask yourself the age-old question of purse thieves:

- Who has something I want badly enough to take a risk?
- Who's slow and weak?
- Who's not going to chase after me?
- Who would you select to be an easy target, a potential victim?

Now, look at your own reflection in a store window, or ask an honest friend:

- How do you carry yourself?
- How are others likely to perceive you?
- Might your overall demeanor benefit from standing straighter, setting shoulders back, and holding your head higher?

ROLE PLAYING: $50 & THE FOOD COURT

While you're at the mall, sit in the food court with friends or family. Pretend each role player desperately needs $50 and has to somehow 'acquire' it from a stranger in that food court (via purse snatching, wallet lifting, smooth talking, or other means).

- Each player scans the crowd for a few minutes, 'assessing all potential victims'.
- Each selects a target. (*You're pretending remember!*)
- Discuss who each player picked out and why.
- What qualities make that person a 'target'?

Now, turn the table:

- Who among you has trait(s) that make you a target? For instance:
- Who has a new cell phone hanging out their back pocket?
- Who wears a huge purse begging to be taken off their shoulder?

If someone grabbed your purse or new merchandise and started running, what would you do? Take off in hot pursuit? Yell to stop him/her?

HOW PREDATORS SEE YOUR HOME

In the same ways that we watch each other, we also look at other peoples' houses.

It's second nature to take note of manicured lawns and perfect porches, or see peeling paint and disarray. Potential predators take in all this and much more. To be an interesting target, your house and neighborhood have to look like they have something, but it doesn't take much. As people become increasingly desperate, pharmaceuticals, a plasma TV, or a wifi tablet can be reason enough to break in, steal stuff, and sell for $20 for their next high. (Given more time, criminals will look for more and take more).

Strangers walking by in the day or at night could be checking out your house. So could your neighbor's kid's friends. It's not uncommon for criminals to pose as badge-wearing technicians, salesmen, or canvassers getting close to your front door for closer inspection. And if you open the door, they'll assess your demeanor and measure your politeness as they look inside at your stuff. (See page 62, Rule 3). Predators locate entries and exits, assess fortifications, weigh strengths and weaknesses, then calculate if the risk warrants the reward. So let's look at what makes a house attractive….[we are, of course, not encouraging illicit behavior!]

ROLE PLAYING: YOU'RE THE BURGLAR ON A SATURDAY NIGHT

Picture yourself as a burglar. You're standing in front of three homes on a Saturday night, each one appears to be quiet. All are nicely built in a good neighborhood. As far as you know, each one has what you want: free stuff.

House #1:   Overgrown shrubs by the windows in the back that would provide cover while you look inside.

House #2:   A fence surrounds it, a few lights are on inside.

House #3:   A fenced-in yard with "No Trespassing" signs, dogs inside, and motion sensor lights are in two places that you can see.

You, the burglar, take in all this information. Assuming that the potential rewards are the same for each house, would you not 'hit' the house with the easiest access and lowest risk? After all, you want to get in and out quickly, with the least chance of getting caught.

Interviews with inmates and community cops tell us repeatedly, that if one house looks 'doable' and another is more challenging, criminals 'hit' the easy one. It's your job to understand what criminals are looking for and reverse engineer your home prevention systems. The irony may occur to you that they are assessing how well you've thought through this process. If your barriers to entry are high, you reduce the odds of being the target of a random crime. Don't make it easy. Bad guys will choose another house in the same neighborhood. It's that simple and unfortunate for someone else.

Preferred homes to plunder have nobody home, easy access, and stuff they can steal and sell. Of note: an equal number of burglaries occur during the day as at night. And a small percent of thieves will even enter a home with people in it—if they think they can get in and out undetected—that's more than a little unnerving. So let's look at things from their perspective…

From the Street, **Criminals are Attracted** to Houses that:

- Have mail and newspapers piled up at the door; delivery stickers on the front door; packages sitting on the front porch.

- Have lots of shrubs and big evergreen to hide behind.

- Leave garage doors open, displaying tools, bikes, and cool stuff.

- Are always quiet (no TV, no music).

- Have no exterior lights, the house and yard are dark.

- Have no window curtains, looking in is easy.

- Display no alarm systems, or signs of home security.

- Show no signs of motion detectors with lights or alarms. (Bad guys hate lights and noise).

You can **Invite Criminals for Closer Inspection,** if you:
- Don't have a fence.

- Don't have dogs.

- Don't have motion sensors lights to light dark entryways.

- Let them see into your ground floor windows.

- Leave doors and windows unlocked, day or night. (One will suffice).

- Leave the attached garage door open, and, leave the door into your house unlocked.

- Have flimsy doors that can be kicked open.

- Have old deadbolts that are easy to pick open or 'bump'.

- Open your door to a stranger—that's the easiest one—be polite and let them look into your house to see your stuff and check you out.

Let's get real—don't do that stuff!

There are many ways to tell if your house is easy prey or a lot of trouble. Similar to how you present yourself in a mall, a residence can and should show signs of strength. Easy targets on the other hand, whether condos in gated communities or mobile homes, stand out by their lack of reinforcements. Ask yourself:

> Know what criminals look for in an easy target,
>
> and work back from that to prevent home break-ins.

- Do I have motion sensor lights outside?

- Are there glass break sensors on my windows? Deadbolts on the door?

- In my apartment, do I have locks or bars on the inside to prevent entry?

- Do I have a peep hole on the front door?

Why the fuss? **Because barriers to entry deter break-ins.**

HALF-BAKED STRATEGIES

Most women are attuned to their personal safety. When we're honest, we know when it's lacking. For instance, too many people, climb into bed at night and lie there with a small resonating fear, taking comfort in a vision of some weird or unsound self-defense strategy. Below are five quotes from women I know.

66 99

"I really don't sleep well at night. My dog is my protection. If he gets up in the middle of the night and moves to that creaky spot in my wood floor, which Beau does most nights, I bolt straight up." – Sally

66 99

"I keep Pixie (a Chihuahua/Rat terrier mix) in my bedroom at night, if she hears something she'll go to the bedroom door and growl. I figure that if someone opens that door, I'm going to swing the alarm clock around my head and hit him with it". – Anne

66 99

"When my boyfriend went away I slept downstairs on the couch. If they're coming, I want to see them. My cell phone's charged and my BB air gun is loaded on the coffee table". – Suzy

66 99

"I've got guns for self defense. I keep them unloaded in my bedroom and keep the bullets on the other side of the house. I know exactly where they are and can get to them quickly...I have grandkids to think about, they're over here a lot". – Maria

66 99

"I have a mock handgun that looks like the real thing. One of my Dad's students made it for an assignment. He held on to it and now I keep it near my bed". – Hannah

Ignoring isn't the same as ignorance, you have to work at it.
- Margaret Atwood

Sally, the first one, lives alone on the edge of a gentrified city neighborhood among poor inner-city residents and drug dealers. The second one, Annette, lives in a bungalow with her dog in an affluent suburb. The third is me, 4 years ago when my then-boyfriend-now-husband left for a long fishing weekend.

The fourth, Maria, lives in a modern house in the country, her closest neighbor 1,000' away. Hannah the last one, lives in a famous historic house with lousy security.

These are the stories we're not talking about, lame self-defense strategies that friends, Moms, aunts, and uncles go to bed with every night. But we're going to change that!

AN EVOLVING QUEEN OF DENIAL

With good information and increasing awareness, an evolving Queen of Denial:

- wakes up to the reality of rising crime,

- takes measures to deter predators and prevent break-ins, and,

- learns home defense, in case her best efforts fail.

As I dropped my old assumptions and tossed worn out stories, space opened up inside. I could see how much time and mental energy I'd spent talking myself out of things that really do exist. Denial thinking had taken up precious brain space. I was willing to let it go.

So I asked myself: What is the opposite of denial? My answer: getting real and grounded and being more present. Then what? Regarding my home, it's about being proactive and recognizing the need for Prevention and Defense.

Though uncertain how to protect myself and prevent harm, my first steps were simple ones:

From the yard, I removed 'weapons' that could be used against us (hatchet by wood pile; sharp gardening gear). I removed a means of access to the second floor (the ladder that had been outside for 3 weeks). I began to consistently lock first floor doors and windows at night. Easy stuff. Over time, with research and dedicated means, we leaned into preventing home break-ins.

We can evade reality, but we cannot evade the consequences of evading reality.

- Ayn Rand

In **Part 2: Prevention**, you'll find **16** home security products, (4 are ready to use out of the box), 3 DIY projects, 3 case studies, and 3 easy things you can do today. Calamity May wants you to put together an integrated and intelligent security system that works for you, your family and your budget.

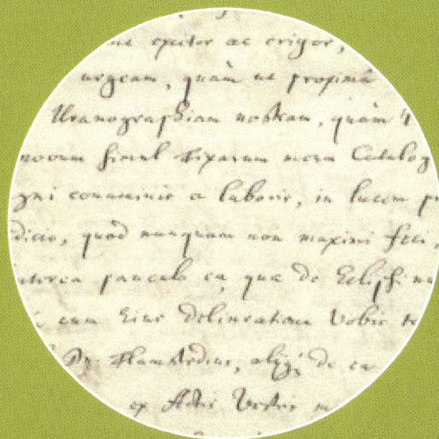

"A word to the wise...
forewarned, is
forearmed."

—Captain Francis Hooke
in a letter from 1685

In the spirit of personal exploration Ralph Waldo Emerson writes "Be an opener of doors".

In the spirit of personal security, Calamity May writes, "Be mindful of securing doors".

PART II

PREVENTION

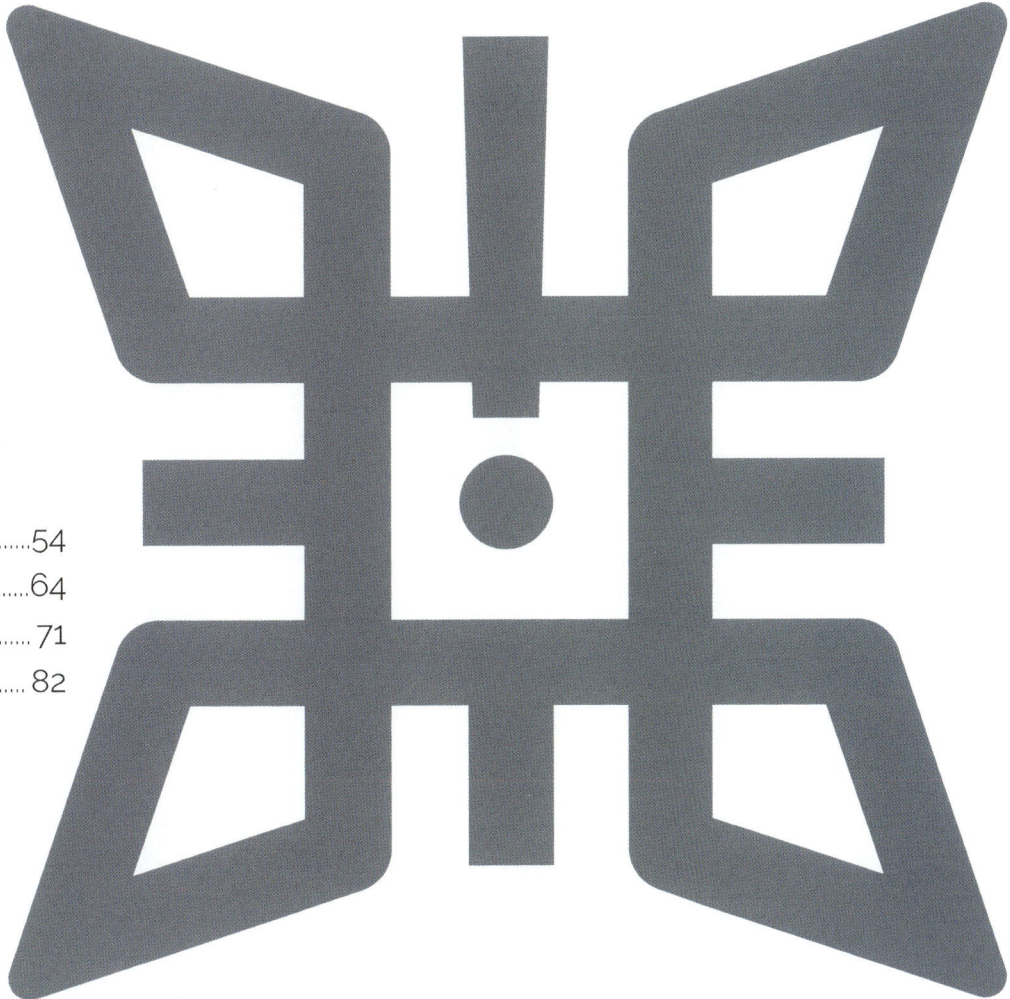

DEFENSIVE CASTLES: A HISTORY

Almost 950 years ago, William the Conqueror invaded and controlled southern England. He and his loyal allies divvied up the new land holdings and built strategically sited Norman-influenced castles for two purposes: to defend their new land, peoples, wealth, and power; and, to serve as safe storehouses and residences. Well-fortified castles thwarted pillagers and stood strong against armies driven by rebellion or political ambition. Strategies employed by castle builders have been at work for thousands of years in forts and fortified cities around the world.

William the Conqueror's defensive castles starting in 1066 were not showplace manors, (those came much later to display wealth and impress visitors), instead, they were simple, functional storehouses and residences designed to detect the approach of war parties and marauders. They had to Eliminate Surprise. To this end, 'keeps' were constructed atop mounds with lookouts on their rooftops to watch for approaching riders or forces, be they friend or foe. From these high points, signals were sent from watchmen. When the alarm sounded, the lord, his villagers, and guardsmen, prepared accordingly.

Early defensive castles were located in places with good sight lines, or on high points from which to survey the landscape. Some were constructed on challenging terrain to make access difficult. When a desirable site lacked defensive advantages to Deny Access, the owner constructed them. Moats were dug, and the keep, each castle's storehouse and residence, was built atop a man-made mound. (A 35' mound with a 3-story tower on top of it was an impressive deterrence in 1070).

A moat is a deep, wide ditch (dry or water-filled). A moat often surrounded a castle to keep out unwanted foot soldiers and horsemen; they also protected castle walls from siege equipment that could be rolled up to their base. Moats focused entry into the stronghold through one controlled access point, typically a drawbridge (that could be retracted). The entry itself was typically a very substantial, thick, wooden door with iron reinforcements that could be barred shut. *These were definitely gated communities!*

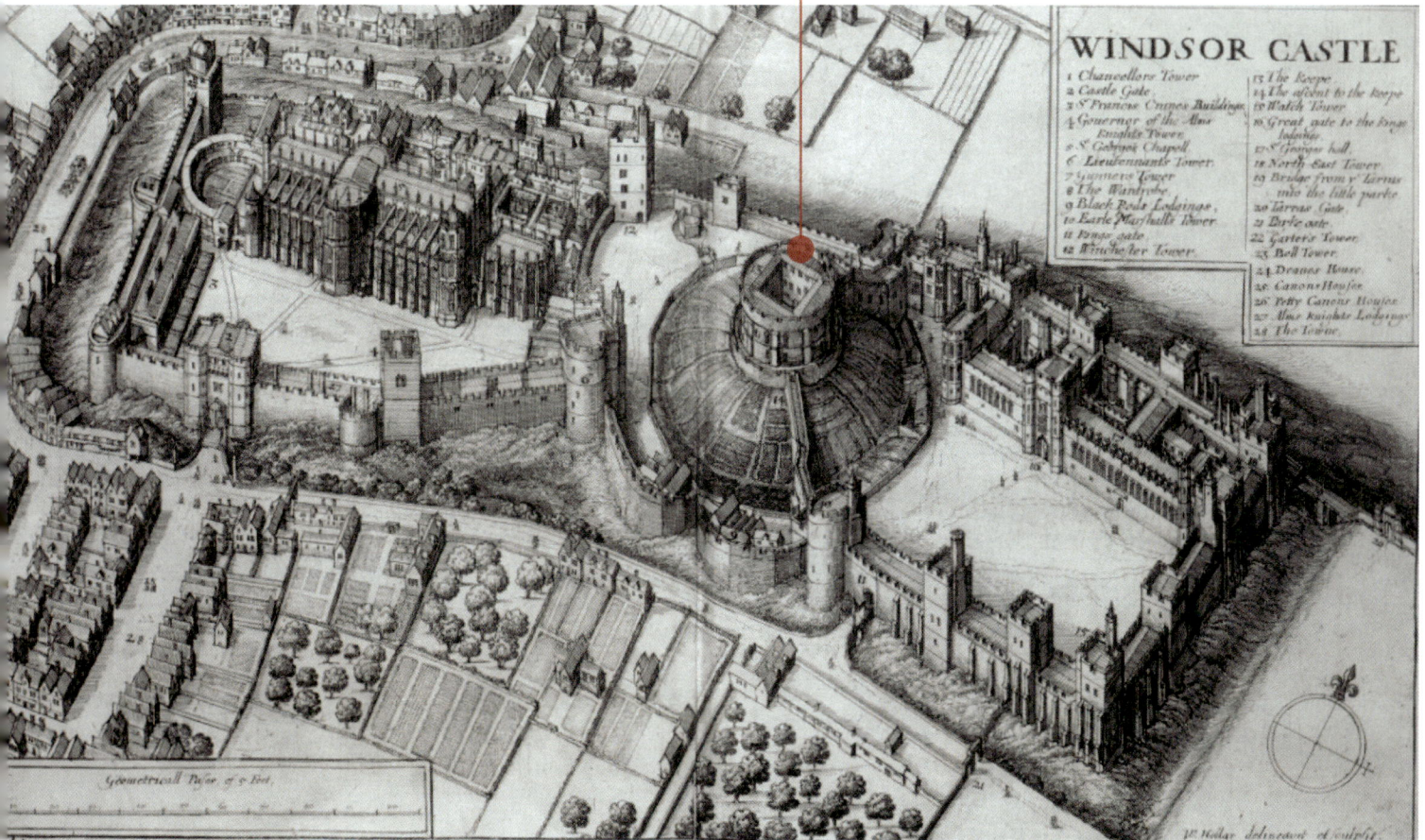

WINDSOR CASTLE

1 Chancellors Tower
2 Castle Gate
3 S.t Francis Cannes Buildings
4 Governor of the Alms Knights Tower
5 S.t Gabriels Chapell
6 Lieutennants Tower
7 Gunners Tower
8 The Wardrobe
9 Black Rods Lodginge
10 Earle Marshalls Tower
11 Kings gate
12 Winchester Tower
13 The Keepe
14 The ascent to the Keepe
15 Watch Tower
16 Great gate to the Kings Lodginge
17 Kings hall
18 North East Tower
19 Bridge from y.e Tarras into the little parke
20 Tarrase Gate
21 Parke gate
22 Garters Tower
23 Bell Tower
24 Deanes House
25 Canons House
26 Petty Canons House
27 Alms Knights Lodginge
28 The Towne

Windsor Castle (above) was constructed by William the Conqueror in 1070. Like all Norman fortifications, at the time of its construction, the outer walls were made of timber. This site in London was chosen for its elevation above the River Thames, and the vast forests and important hunting grounds adjacent to it. The prominent round building on top of a circular mound is the keep (red dot), the castle's earliest structure. Windsor has been occupied continuously for 945 years. Illustration from 17th century, source unknown.

In early Norman castles, to get close enough to even ponder entry to such a place was to be hailed upon by archers' arrows shot from narrow slots on high.

The heart and center of every castle was its keep. Keeps were taller than the castle walls and they served three critical purposes: watchtower; storage for munitions, food, water, wine, wealth, booty and live animals, (and sometimes prisoners); and, residence to the nobleman and his lady. Under attack, the keep was the last refuge and final defensive position. It was crucial to deny access to it and defend it intensely. All keeps were built with extra reinforcements, some with extraordinary measures to deny access.

When defensive efforts inside the castle or keep looked grim, the man or woman in charge knew to Have A Plan!, an exit strategy. Many early castles had built within their walls a concealed exit, or 'postern' gate, where nobility escaped through pre-constructed tunnels, or in some cases, sewers, to emerge far from their castle under siege, where things ended very badly for those left inside. Prevention was so critical, that the bulk of many defensive castles stand to this day, their foundations, at least, can be traced in many pastoral fields.

Suffice to say, well protected and defended complexes demanded too much time, exposure, energy, and risk for would-be intruders. Pillagers who knew of such fortifications looked for easier targets elsewhere.

PREVENTION DEFINED

Today's Windsor Castle from the Main Gate. The keep, to the far left, remains the tallest building in the compound.

Merriam-Webster's dictionary has a simple, almost profound definition of prevention that is relevant to many aspects of our lives. I find it particularly well-suited to preventing home break-ins:

1) to be in readiness;

2) to deprive of power or hope of succeeding;

3) to keep from happening or existing.

By tweaking this definition we tap into very old knowledge, the design and defense of the earliest castles. *Lo!* if you live not in a castle, but in an aparptment, a modest hut, yurt, or postage-stamp walkup—I say to Ye: It is not the address nor the expense of your walls, it's what's inside! You. Your kids. Fido. Mother. Aunt Favorite, and the booty she plundered over the years. The castle metaphor gives us a useful message from a long past world without wifi, phones, or 9-1-1. It reads: Ye shall put into place lots of prevention before resorting to the dire need for defense. Like your ancestors long before you, do your very best to prevent intrusion.

One thousand years later, that hasn't changed. Calamity May presents a twist on Webster's definition to prevent a breach of your castle:

ZONE 1: ELIMINATE SURPRISE (deprive the bad guy of power and hope)

ZONE 2: DENY ACCESS (make entry difficult & time consuming)

ZONE 3: HAVE A PLAN! (be ready in case of a breach)

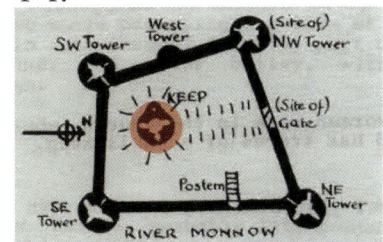

It bears repeating: wherever you live, your residence is the keep–the safe haven–if you make it one. Above, a sketch of an early Norman fortification shows five key elements:
· Keep (red)
· Watchtowers on corners
· Walls
· Gated entry
· Escape route (postern gate)

The Watchtower:
Exterior Detection Systems

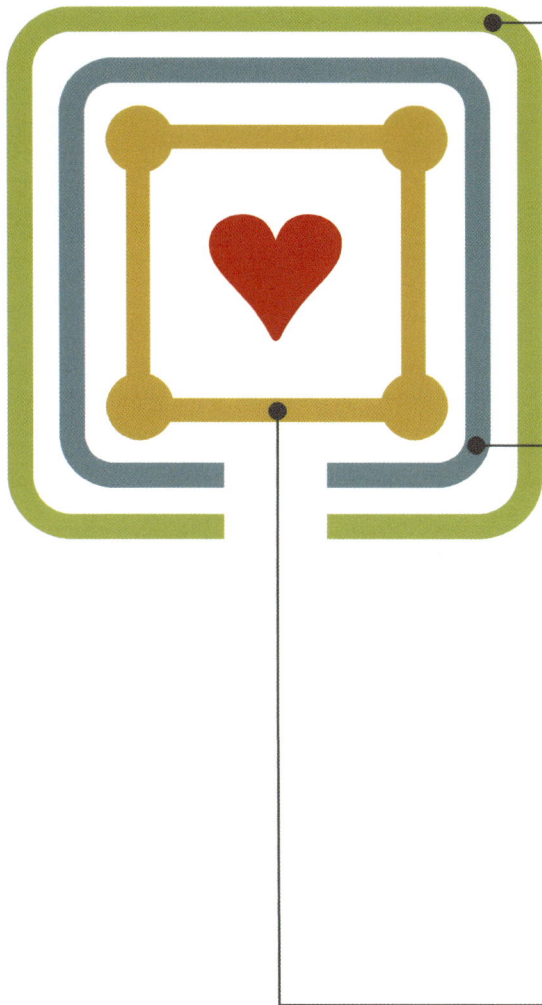

Zone 1

The Moat:
Property Boundaries, Yard, Hallway
Eliminate Surprise
Deprive the bad guy of power and hope.

- Driveway sensors
- Motion sensors with alerts, lights, camera or video
- No Trespassing sign
- Inside lights & TVs on timers
- Dogs

Zone 2

The Drawbridge:
Doors & Windows
Deny Access
Thwart intruders: make entry difficult & time consuming.

- Lock all doors & windows
- Do not open your door to strangers
- Upgrade deadbolts & window locks
- Bar doors from inside
- Cover glass in doors with non-breakable materials
- Use psychological deterrents
- Train your dog to bark on demand

Zone 3

The Castle Keep:
Inside Your Home
Have a Plan!
Be ready in case of a breach.

- Have an escape route
- Conceal yourself
- Have self defense tools in every room
- Retreat to a safe room
- Know your defensible positions
- Have cellphone and weapon ready
- Practice with the whole family!

ZONE 1: PROPERTY BOUNDARIES, YARD, CORRIDOR

ELIMINATE SURPRISE

Deprive the bad guy of power and hope.

UNDERSTAND YOUR NEEDS

Identify threats that actually exist. Talk to neighbors who read the paper, or watch your street like a hawk, or listen to police scanners. Look online to see if local police publish an annual police report. (The Pittsburgh Bureau of Police publishes one that's very detailed and informative). Talk to the sergeant at your local police station, ask about recent crime trends and break ins. Inquire about your street and those adjacent to you. Ask where the hotspots are. (Don't be intimidated, they work for you).

KNOW YOUR BOUNDARIES

To eliminate surprise, be forewarned of an approaching person or persons. Watchmen of yore have been replaced with small devices sensitive to movement, body heat, or metals, that when triggered, can turn on lights, make noise, take photographs, or start videotapes. Zone 1 is where you install these things. The outer edge of Zone 1 may be your property boundary; it may be the hallway outside your apartment. Whether you live on 1/8 of an acre, or 80 acres, Zone 1 extends a reasonable distance from your house, not necessarily to the full perimeter. Assess where and how someone can gain access to your property and work back from that. If you're alerted, have a plan to respond accordingly.

KNOW THEIR TRICKS

Know what bad guys look for! If they find your home a possible target, they'll get close enough to look in your windows and doors. They'll assess valuables and people inside, (to prey upon or contend with). They want in and out quickly. But first, they have to figure out which is greater: the risks or the rewards.

Here's the fun part: Let's size up your house, see who's inside, see if there's stuff worth stealing, and determine if the effort's worth $50. Get up, put shoes on, walk around, and consider these points:

- From inside: Look out each window and door, especially those on the first floor. Locate places outside where a person can see in.

- From outside: See the views into your house from the yard or sidewalk. What rooms can you look into? Kitchen? Living room? Which parts of those rooms are visible? *C'mon now...get up and look!*

- What stealable stuff or vulnerable person(s) do you see? Do this in daytime and at night. (Predators do).

- Where would you have the most time and least detection to break into your home. The back porch by the overgrown hedge? The basement door?

Now, step back to the edges of your property or apartment.

- How would you approach your dwelling? Through a neighbor's yard? The woods? The fire escape?

- Where would you hide until all or most family member have left? Behind a bush?

Use this process to help you locate where outside detection devices makes sense, before unwelcome eyes get close enough to peer into your doors and windows.

BE STRATEGIC

Do your best to deprive the bad guy of "power and hope." This is your castle, you're the keep—the heart and soul of it. Don't let them close in without detection. Change the minds of pillagers and predators while they're outside. Prevention is crucial.

If the idea of being alerted to a trespasser outside your home or apartment balcony gives you the creeps; if you're reading this because you live alone and/or live in a declining neighborhood, then keeping trespassers and criminals out of your dwelling is imperative.

Think strategically about Zones 1 & 2, and work toward an integrated security measures. Adopt a mindset that's proactive and aware.

What follows is relevant info for all ages! By the end, you and your teenager will have a better idea how to implement strategies to prevent home break ins. Many products are inexpensive and easy to install.

REALITY CHECK:

A determined intruder with enough time and proper tools can get in to any home.

CALAMITY MAY'S RESPONSE:

Vastly improve your odds

Eliminate Surprise · Deny Access · Have a Plan!

ZONE 1 CATALOG:

HOW TO ELIMINATE SURPRISE

The Zone 1 Catalog presents ideas for home security strategies and products worth implementing. Whether you live in a 3-story walk-up, or a 20-acre estate there are products and practices that can be tailored to where and how you live.

Zone 1 extends to the property boundaries outside your home. In apartments, it may extend to the hallway, balcony or fire escape. Zone 1 is where you detect and eliminate the surprise of an intruder closing in to where you live. How?

DETECTION

Outdoor sensors are your watchmen and sentinels. Sensors detect movement and transmit signals to a receiver (or base station) inside the house. The receiver inside, emits a sound to alert you. The receiver to the right (page 55), can pick up signals from up to 4 outdoor sensors.

VISIBILITY

Always look through your door before you open it. Get the full picture first!

SIGNS

Signs can have power if you back them up. Private Property, Beware of Dog, home security signs can't be stand alones though. If an intruder tests your system and finds no enforcement, you've created a false security layer, and they know it.

The purpose of Zone 1 tools is to detect movement. In the place of a castle watchman, they let you know when someone's on your property, before they get to your doors and windows.

Four sensors on the following page are from Dakota Alert. They sell separate components so you can build a customized, integrated system. They have positive consumer reviews, helpful customer service, and prompt delivery.

PERIMETER

No Trespassing Sign

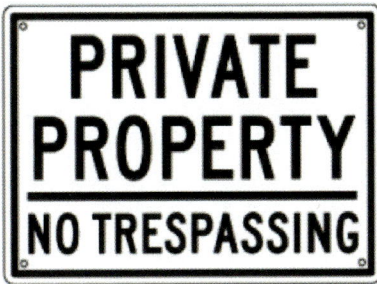

The simple No Trespassing sign. If your property borders a road, alley, trail, footpath or woods, consider posting a Private Property No Trespassing sign. It's legal notice, and sends the message: "You're Not Welcome and You've Been Legally Notified".

Example: If you want to be legally defensible, put a No Trespassing sign in the background of a motion-sensitive video camera. If a bad guy peers in your windows, or tries to get in, you have him on video with a No Trespassing sign behind him.

It may be useful in front of the magistrate.

Dakota Alert Receiver

This is Dakota Alert's receiver that works with the 4 sensors on pages 56-57. It rings a chime inside your house (or workshop) to let you know something's been detected. One receiver takes input from up to 4 sensors. You choose a unique sound to assign to each sensor, this lets you identify where motion was detected in your system. Example: One receiver works with

up to 4 different sensors. The varied sensors may include:

- a wireless driveway sensor,
- (2) 300-foot long IR beams along your property,
- a motion sensor for on a garage or outbuilding.

DCR-2500, pair with sensors, shown on following pages.

DRIVEWAY

Wireless Driveway Alarm

Detects vehicles only

Reliable driveway sensors use magnets to detect steel in vehicles. The magnetic sensor is buried parallel to the driveway. When a car passes within 10-12 feet of it, the magnet emits a signal that's sent 1/2 mile to the receiver in the home. The receiver (page 55) then chimes and/or flashes. This kit comes with: the probe, transmitter, 50 feet of wire (from probe to transmitter) and a receiver (placed in the house or workshop). Magnetic driveway sensors do not detect animals or people.

DCPA-2500, $319.99, sensor and receiver.

DRIVEWAY & ANYWHERE

Wireless Motion Alert

Detects cars, people, animals

BUILD A BIRDHOUSE AROUND IT !

This outdoor, infrared sensor detects anything that generates heat and crosses its invisible, infrared beam: a warm car, a person, or a bird flying up to 50 feet in front of it. The built-in transmitter sends a signal up to 1/2-mile to the receiver. The sensor needs to be placed 20'-30' away from the driveway, and 3' high to detect cars. Installation is easy. Follow directions to avoid false alarms, or no alarms. It will detect the body heat of animals (like deer) at 3' or higher.

DCMA-2500, $159.99 sensor and receiver. Sensor only, $89.99

DIY: Build a Birdhouse for Your Sensors!

Create a cover for your sensor and mount it on a pole or post without being obvious.
DIY Notes: Take measurements of your sensor to make sure the opening in the bird house aligns with the lens' line of sight. Keep birds out by using a dense enough wire mesh to obstruct entry, while maintaining a clear view. for the sensor.

PROPERTY PERIMETER

300-foot long Wireless Break Beam

Detects movement

This infrared detector's good for larger properties or those with one or more long boundaries. It would be useful on properties next to trails or alleys, for instance.

This wireless, solar-powered unit is simple to install, no wiring needed. The sensors can be placed up to 300' apart, and send a signal to a receiver up to ½-mile away. When the straight infrared beam is broken by a person or animal (at the height of the the infrared beam), a signal is transmitted to the receiver (in the house or workshop). One receiver (page 55) can take in input from 4 pairs of sensors.

BBA-2500, $329.99 pair of sensors and receiver. Sensor only,$259.99

SMALL, PROTECTED SPACES

Motion Sensor for Smaller Spaces

Detects movement

This small passive, infrared unit detects movement indoors, or outdoors if protected from rain (under a porch roof, for instance). It's highly adjustable in three ways: 1) it has a cone of detection that adjusts from 20 to 100 degrees wide; 2) it swivels left and right; and, 3) it swivels up and down. It has three sensitivity settings that allows it to detect movement at 12, 25, or 40 feet. When it detects motion, an LED flash can go off, or, the unit can stay dark as it transmits its signal to the receiver. Works with receiver, page 55.

IRDCR-2500, $129.99 sensor and receiver. Sensor only, $84.99

PROTECTED SPACES

Motion Sensor and Digital Recorder

Records Video

USE WITH NO TRESPASSING SIGN IN BACKGROUND

This tool uses infrared detection and a digital video recorder to record 30 seconds of video and audio every time a person passes in front of it. It's designed for indoor use, but can be used in a protected outdoor space sheltered from rain, snow or fog.

Video is recorded on a 4GB SD card. It's battery powered, easy to set up and use, and works in low light conditions (it doesn't need light to take images).

The camera / video connects to a USB cable, and can be viewed on most computers. The lithium battery lasts 10 days.
DVR-01, $129.99

NOTE: Sensors based on infrared detection do not work through glass. An IR-based sensor placed on one side of a window will not detect body heat of persons or animals on the other side of that window.

DOORS, STAIRS

Wireless Floor Mat

Detects weight

Dakota Alert's wireless floor mat's designed to be placed under a rug or welcome mat so it's not noticeable. When someone steps on it, a signal is sent to a receiver inside the home or apartment. The receiver sounds a chime and/or flashes a bright light to notify that someone's at your door.

The mat and receiver both run on batteries that come with it. (The batteries that came with mine had expired, the mat didn't work. But it was an easy $2.50 fix for new batteries). This sensor also serves as a doorbell to let you know when company arrives.
MSRP $129.99, sensor and receiver

I really like this product. Read my review at calamitymay.com, "Turn Your Welcome Mat Into a Lighthouse".

NOTES ON SENSORS

Unless solar-powered or hard-wired, outdoor sensors run on batteries. Keep them fresh.
Many sensors, depending on location and height may detect the movement of animals. Proper installation can prevent false alarms. It's smart to think in terms of a 'layered system', where sensors are located on an outer perimeter **and** up closer to your home. You got it, Zones 1 and 2.

Call Dakota Alert to discuss your needs, let them suggest your best options. The descriptions on DakotaAlert.com are detailed, but you'll learn more by talking to them. (*I did!*)

DOORS

Digital Door Viewer

Doorbell, Camera, Video

With Mul-T-Lock's digital door viewer you can see who's standing outside your door and a lot more.

Mul-T-Lock's GotU+ Digital Door Viewer with Bell is an advanced digital door viewer. When someone outside touches the doorbell, from inside your house, you can:
• see who's at the front door
• have photos or 20-second video taken of the person at the door
• see images in low-light or night
• see date and time stamp on images
• playback still images or videos using its internal memory.

Useful for homeowners and shopkeepers.

Mul-T-Lock products are considered top of the line. The viewfinder (peep hole) installs in the door with an electric drill; setup is said to be straightforward. This model comes with a doorbell touch pad. ACC-DDV. MSRP, $199.

Model GotU+ 5140 does all the above, but its motion sensor triggers video taping and still images and sends a chime automatically **before** they reach the door. MSRP, $219.

DOORS

290-degree Door Viewer Scope

USE WITH ZONE 1 WIRELESS FLOOR MAT

New Vue Trading

If you can't see through your door install a peephole! Better yet, install this one with a wide-angle lens that lets you see a visitor's face, a small package sitting against the door, or the car parked across the street.
The New Vue Trading door viewer has a unique 290-degree viewing radius. It's made with 4 high-quality glass lenses, costs $5, and is easy to install with an electric drill. (It needs a 5/8" hole). From start to finish, installing one may take 15 minutes.
New Vue Trading, MSRP $4.99

Door viewers with narrow fields of view miss important information, demonstrated in this example:

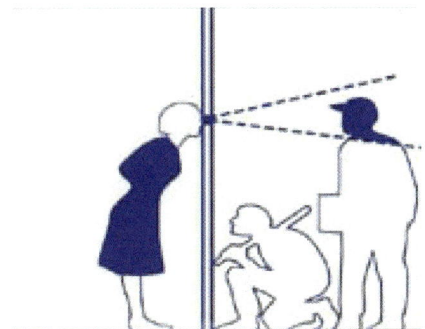

ALWAYS look to see who's outside before opening your door. NEVER open your door to a stranger.

ZONE 2: DOORS & WINDOWS HOW TO DENY ACCESS

Thwart intruders: make entry difficult & time consuming.

For a bad guy to get into a home, he/she has to walk through a door or climb into a window. Your job is to make that difficult. Let's review some facts, and rules, attend door school, and enter the product catalog.

Doors allow us to enter and leave dwellings. They can be prominent and strong, or weak with wear and abuse. We rarely think of doors anymore as historic lines of defense, but they are.

Windows make homes livable. They let in light, fresh air, and allow views to the outside. It's ironic that architects refer to doors and windows as 'openings', because both pose the foremost security risks to breaking into peoples' homes. In the U.S., two-thirds of all completed burglaries, allowed access through unlocked doors or windows.[1] The bad guy walked in, or climbed in a window. The other third used force to gain entry.

"Forcible entry" uses physical force, a prying tool, or lock picking to gain entry. Both types of entry are, of course, illegal. Felonies, in fact.

Once in, the next step (typically) is to steal stuff. It's called 'theft', taking something that is owned, or in the proper custody of another person. There are two types of theft: Burglary is theft without harm. Robbery is theft by instilling fear or using force or violence.

The most common types of forcible entry are, in **descending** order:

1. Kicking in a door
2. Breaking glass in the sidelights (adjacent to the door), then unlocking the deadbolt
3. Prying a door open at the lock
4. Picking or bumping the lock
5. Lifting a sliding glass door from its tracks

According to Jack McLean's 1983 autobiography *Secrets of a SuperThief*, (and the survey he conducted while in prison), 41% of burglars will go into your house while you're home, if, they think they can get in and out without detection. (*What!?*) You **do not** want this to happen to you, or the baby sitter and your kids. Twenty-eight percent of home burglaries have someone home, one-quarter of those people experience violence from the intruder.[2] Remember what Brian Keith said (page 26): "It gets ugly if somebody's home."

Getting inside a home is easy because most people don't follow a few tried and true rules. Below, are The Big 3, and how Joe & I responded accordingly.

THE BIG THREE

1 Rule #1: Lock your doors and windows.

Fact: Burglars, armed robbers, rapists, and deranged -exes enter homes most frequently through unlocked doors and windows.

What we did: Our doors got a lot attention, they needed it. (See pages 71, 76, 77). Over a 2-year period Joe and I installed the following products that work with our lifestyle.

Curtains. For years we went curtain free on the first floor. Loved it. Then I learned about prying eye potential, and we put up opaque linen curtains. We close them consistently when we go out, and before going to bed.

Movement Sensors. Movement sensors on a window detect if a window moves up or down based on a shift in magnets. If there's movement, a signal is sent to a receiver upstairs, where a chime sounds. We installed sensors on each first-floor window that has access from our porches.

Glass Break Sensors. Glass break sensors detect the unique sound of breaking glass. Each sensor covers a 20-foot radius. We installed one in our first-floor.

Pin locks. Pin locks allow us to leave the upper sash windows open to 4" (to let hot air out) and lock it into place. [Why a 4" maximum opening, you ask? Sometimes burglars use little kids to gain access through small openings to get into a house and unlock the doors. Four inches denies access to small heads].

Window latches. In renovating our old house, we installed new double-hung windows throughout. New windows mean new window locks. We use them consistently.

Result: 5 "layers" of window security.

2 **Rule #2: Affect a criminal's perception by instilling doubt into his/her mind. Make your dwelling a time and energy sink, and they'll move on.**

Fact: Make your home hard to break into and the bad guy will go to another house.

What we did: Some of our our Zone 2 tools are noticeable, others aren't. They include: the aforementioned window security, a home security system (sensors & receivers), security sticker in a window, motion detector lights, new deadbolts, barred doors, and big dogs.

That's a layered system that would take a bad guy a while to get into–if he's determined. Most wouldn't bother. It would take too much energy, time, tools (and noise) to break in unnoticed. Not to mention, the Giant Schnauzer would be a pretty ticked off about the unscheduled visit.

3 **Rule #3: Never open your door to someone you don't know, did not invite, or did not call over to your house. Especially critical for the young and elderly.**

Fact: 80% of crimes against property and people are committed by people you know or have met, even if briefly.

What we did:

1) We rarely open the door to strangers. We're now even cautious with "professionals".

2) Opening your door to someone who looks like a technician or a canvasser, or respectable-looking professional is an easy way for a bad guy to get into a nice house on a well-watched street. It's an old trick, don't be fooled.

3) If the guy who installed your new Internet router, or, if the mover who brought in the fridge shows up later because he forgot something–well, have that conversation with the door closed between you. Tell him you'll send it to his manager if you find it.

4) Furthermore, never open your door to:

• a distressed person who claims they need to use your phone. (With the door closed, offer to make the call for them).

• a person claiming to be a friend of a friend. (Call to verify).

!

Tough Rule

This rule is hard for many of us to follow. However, if you are vulnerable, handicapped, disabled, or elderly, or, if you're a teen-age babysit-ters, it makes sense.

Honestly, I struggle with this one. I grew up in the spirit of hospitality and kindness to strangers. But more often than not, I now keep our front door closed to salesmen, canvassers and religious types. When I do open it, the dogs are right there. But I imagine that someone in need, or someone who may be a neighbor I don't know yet–that's hard. I wish we had a peep hole (the one on page 59), so I could see who's there. I'll put that on the Fortify Castle list.

REVERSE ENGINEERING

Men, women and juveniles who are in jail right now (or not, remember unreported crimes) are swapping stories about illegal break-in strategies that work or fail. Let's make it our topic too. Real (not assumed) security of your doors and windows calls for a study in reverse engineering. Learn to identify your weak points and fix them. You are about to enter a catalog of tools and techniques to get you started in the right direction. But first, learn why doors are so vulnerable to illegal entries.

Welcome, to the Door Education section. Come in, it's not locked...

What's super simple, has proven to work for thousands of years, and costs $20 to make?

The barred door.
See page 75

In 1983, Jack MacLean wrote his autobiography while in jail. It's called *Secrets of a Superthief*. It's a fascinating read to gain insight into the workings of a highly successful (so he claims) high-end residential thief. He says he wrote his book to help the likes of us homeowners, and he does offer good advice I've not seen elsewhere. His book includes a survey of 129 questions he posed to 300 fellow inmates who were in for similar felonies of larceny and theft. This survey is considered unique and useful for its lack of academic bias. Below, is an example. He wrote the answer after compiling all the responses.

Q: *"Which would you go through more often, doors or windows?"*

A: "Naturally all burglars have gone through both in their days, but the question asks which they went through most. Seventy percent stated that they went through more windows than doors. The other 30 percent were like me—going in on their own two feet through the door. This shows a great weakness in windows and their locking".

DOORS: ANATOMY & FUNCTION

Doors. You go through dozens every day. They're part of every building's security, privacy and weather proofing. They're obvious necessities. Unfortunately, exterior doors, by their design, can be vulnerable to illegal entry.

RECOGNIZE VULNERABILITY

Noblemen spent great sums of money on outermost doors to their defensive castles, they had to. Yours are equally important. There are good ways to strenthen exterior doors, and ways to recognize when one needs to be replaced.

It never occurred to me to understand how a door works, let alone how to assess the security of one. I finally learned though, after much head banging that DOORS ARE EASY TO ENTER and DIFFICULT TO SECURE. My best friend/husband/architect patiently explained a door's many parts and drew diagrams. Here, I share with you what I learned (minus the head banging).

In this section, I lay out:

- the structure of a door, its hardware, and how they work together

- list 10 points that make doors vulnerable and how you can fix them

- show products and ideas to secure doors, and

- share 4 case studies from our home

When you learn these basics, you'll be able to assess the safety of your own doors. But first, there's a significant limitation within the design of doors. Understand this, and the rest falls into place....

Force on a Door

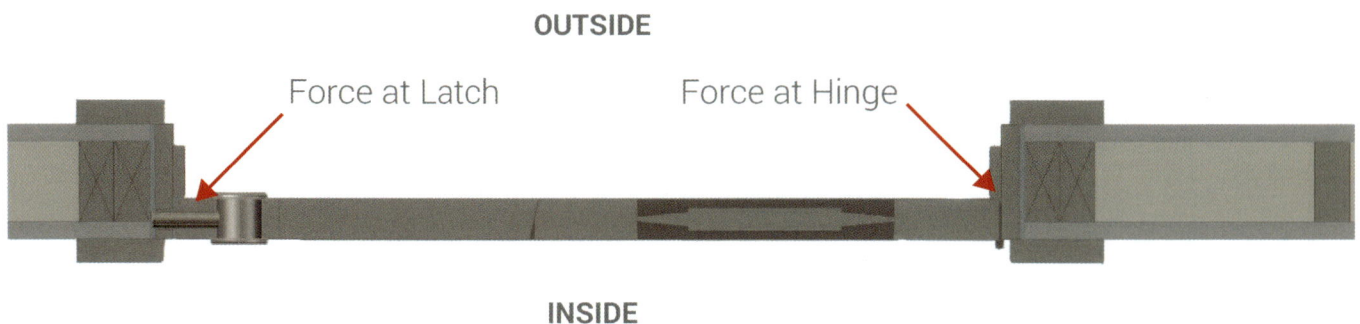

OUTSIDE

Force at Latch   Force at Hinge

INSIDE

THE FLAW. Doors that swing open hang on the outermost edge of the door frame.

Most exterior doors open to the inside. For a door to swing in 180 degrees, 2 things have to happen: the door has to attach to the outermost edge of the frame; and, the hinges have to jut out between the door and the casing. This positions a door to sit on the very edge of the door frame, relying solely on its hardware to attach it to the structural frame.

While it's fresh in your brain, get up and look at your exterior door. See where the hinges are? See how the door sits on the very edge of the doorway? What condition is the jamb in? (See definition on page 67). Is it undamaged or has it seen a pry bar or a hefty kick? What condition is the wood around the handle/latch and deadbolt? Undamaged or beat up?

Why are doors so vulnerable to forced entry? Strong kicks at the hinge or door handle can bust a door loose if not firmly attached. If the jamb is weak in any way, or if screws that are supposed to drive into the frame are too short, the security of your door is compromised.

The good news: these critical points can be reinforced with light to modest effort!

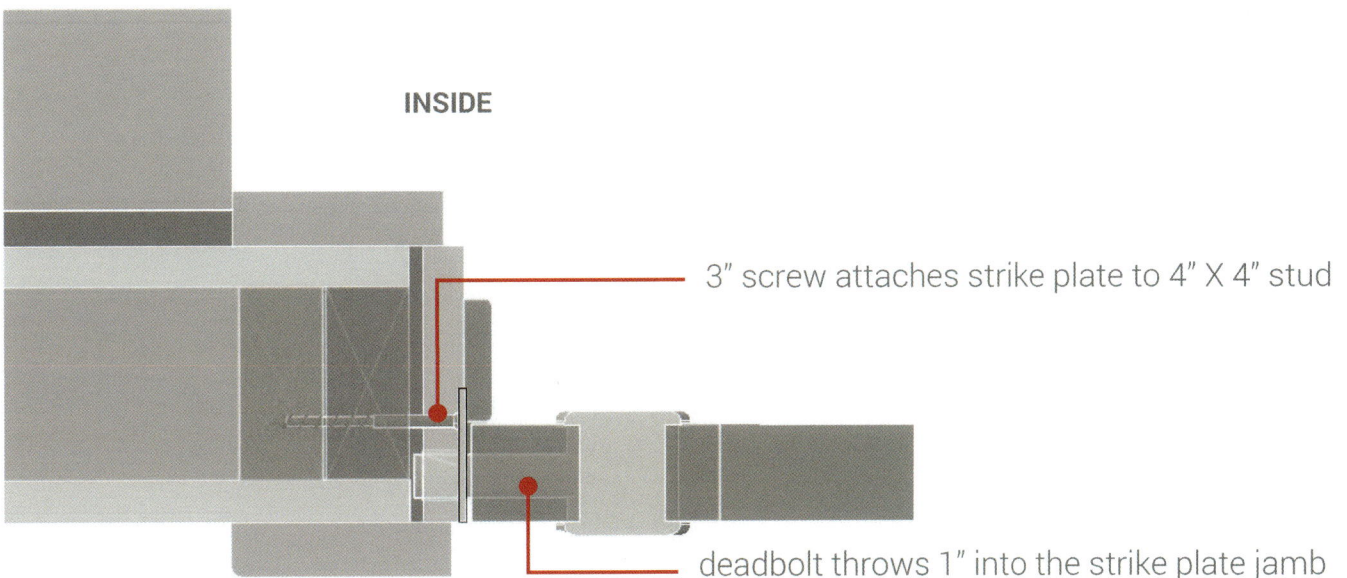

INSIDE

3" screw attaches strike plate to 4" X 4" stud

deadbolt throws 1" into the strike plate jamb

Anatomy of a Door

structural door frame:

lintel (top)

double stud posts (sides)

studs, 2"x4" pine boards every 16" (typ.)

drywall infill (unless brick or stone)

hinges: 2 or 3 per door

door handle & latch

deadbolt

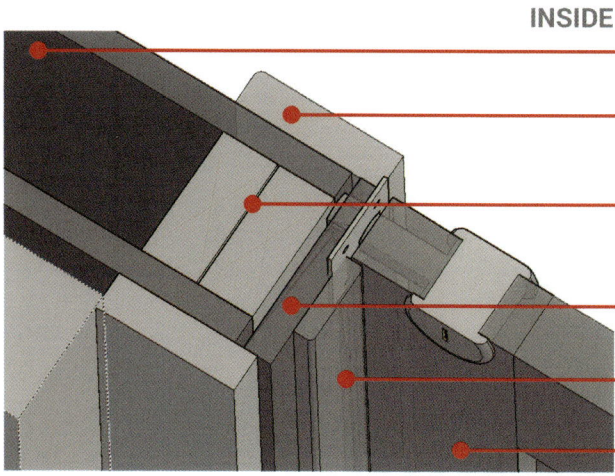

INSIDE

drywall infill

casing (or trim)

structural door frame:
double stud post

jamb

stop

door

OUTSIDE

Looking at an exterior door, you see a door, a handle, a deadbolt (typically), and maybe a window in the door. The door itself is framed by, and attaches to, a structural frame behind the finished wall. It's the backbone of the door system. (You'd have to visit a house under construction to see how it's all put together). For exterior doors to be strong and secure, they must strongly attach to this frame. This is the critical point to making exterior doors secure as possible.

ANATOMY. Whether your home or townhouse is built of brick, stone or wood, almost every **door opening** is framed in wood.

An exterior doorway starts out as a frame built into a wall. Its two vertical sides are sturdy 4" x 4" posts, **doubled up 2" x 4" studs.**

A wood **lintel** spans the top of the doorway and screws into the posts.

The **jamb**, (the finished wood lining) covers the frame and attaches to it. All door hardware attaches to the jamb. For real door security, 3 parts should screw through the jamb and into the 4" x 4" posts:
- Hinge screws
- Strike plate screws
- Steel door armoring screws (page 74)

What screws to the jamb only?
- Door **stop**

Casing is the finished trim you see inside a room, and on the outside of a door. It finishes the rough edges of the frame and drywall or plaster.

Glossary

casing: the finished trim that frames a door. The casing hides rough edges and provides moderate to significant strength.

deadbolt: a lock that projects a steel bolt from the door into a metal strike plate. Doors have one or more deadbolts. Deadbolts are stronger than locks in the door handle.

handle & latch: the handle is what you lay your hand on to open a door. It releases the latch which keeps the door closed.

hinge: hinges have 2 distinct parts: the round metal cylinders on which a door rotates, and, the hinge plates that attach to both the door and the jamb (and screws into the 4" X 4" post).

jamb: finished wood that "lines" the inside of a door frame and attaches to the structural frame. Hinges and strike plates need to penetrate the jamb and screw into the 4" X 4" posts behind it.

latch: the metal protrusion that holds a door shut. A door's handle moves the latch.

stop: the vertical wood piece that stops the swing of a door. It's what a closed door rests against. A stop screws to the jamb.

strike plate: the metal hardware that receives the latch and/or deadbolt. It screws into the jamb (and the 4" X 4" post behind it).

structural door frame: the overall assembly onto which a door hangs. It's comprised of a post on each side, a lintel, a jamb. All of which can only be seen during construction.

10

10 Vulnerabilities of Exterior Doors

Words in bold are explained in the Glossary.

PROBLEM	SOLUTION
1. HABIT A door is left unlocked. (The #1 means of illegal entry).	Lock all exterior doors, when you're home and when you leave.
2. HINGE SCREWS Short screws (less than 3") attach **hinge plates** to the **jamb and frame.**	Take out 1 screw from the hinge plates that attach the door to the frame. Measure it. If it's less than 3 inches, replace 2 screws per hinge plate with 3" steel, drywall screws. Preferably stainless steel.
3. STRIKE PLATE Short screws (less than 3") attach the **strike plate** to the **jamb and frame.**	Take out 1 screw from the strike plate. Measure it. If less than 3 inches, re-place it with a 3" steel, drywall screw.
4. DEADBOLT Short **deadbolts** throw less than 1" into the strike plate	Turn your deadbolt so it sticks out. Measure its length. If it's less than 1" inch long, replace it with new deadbolt. (Zone 2 catalog). Install new stike plates for new, longer deadbolts.
5. OLD LOCKS Old or low-grade locks can be readily picked, bumped, or pried open.	a. If your handset and latch is the only lock on the door (no deadbolt), buy and install an ANSI Grade 1 deadbolt and strike plate. (See Zone 2 catalog). b. If your deadbolt is old, or wobbly, replace it with same. (Zone 2 catalog).

PROBLEM

SOLUTION

6. GLASS PANES IN DOORS

Glass in the door panel can be broken, followed by a hand reaching in to undo deadbolt and latch.

An exterior door with lots of glass has 2 options:
a. Replace the door. Windows are fine at the top of a door, where breaking and reaching in would be difficult. (Case Study 2).
b. Protect the glass. (Case Study 4).

7. GLASS SIDE PANELS

Glass side panels can be broken, a hand reached in to undo deadbolt and latch.

a. Fill in glass side panels
b. Install steel screens inside or outside.
c. Protect the glass. (Case Study 3).

Glass side panels separate a door from its structural frame, creating a zone of weakness.

a. See above. Good (and creative) carpentry techniques could reinforce the door-panel-frame attachment.
 (Case Study 3).

8. WEAK AREA AT HANDLE & LATCH

Door material is removed to carve out room for door **handles** and **deadbolts**, weakening those areas.

a. A door's handle/latch area can be reinforced using steel reinforcement, (See EZ Armor, page 74).
b. Install a new, sturdier door.

9.AGING DOORS

Doors age and deteriorate when exposed to the elements.

a. Install a new door, (Appendix G), or, a used, robust doo).
b. Reinforce both sides with steel reinforcement (page 74).

10. DOORS THAT OPEN OUTWARD

An outward-opening door can be taken off its hinges

Install hinges with non-removeable pins,

THINGS TO DO TODAY

Tools you'll need:
- A measuring tape
- A screwdriver (preferably, an electric drill)

1 HINGE SCREWS

Action:

Take out 1 screw from the hinge plates that attach the door to the frame.
Measure it. If less than 3 inches, replace 2 per plate.

The Fix:

Replace 2 hinge screws. per hinge plate, with 3" stainless steel, drywall screws.

2 STRIKE PLATE SCREWS

Action:

Take out 1 screw from the strike plate.
Measure its length. If less than 3 inches, replace it.

The Fix:

Replace 1 strike plate screw with 3" stainless steel, drywall screws.

3 DEADBOLT

Action:

Turn your deadbolt so it sticks out. Measure its length. If it's less than 1" inch long, replace it.

The Fix:

Buy a new ANSI Grade 1 deadbolt and have it installed, or do it yourself (moderate skill needed). You may need a new strike plate to accommodate a longer deadbolt.

CASE STUDY 1:
ASSESSING WEAK AREAS

REAR WINDOW DOOR

Two hinge sets attach to the door to the jamb with 1-1/2" screws. Way too short!

Several years ago, we turned a solid wood door into a window door for the view. We used tempered glass which is harder to break, but, it can break. Also, the expanse of glass weakens a door's structure.

Doors are hollowed out to accomadate latches and deadbolts, weakening that area, especially in older wood doors.

Age, sun exposure and lack of maintenance over decades, have rendered the panels in this pine door brittle. Oiling could help preserve the wood's integrity.

Our Solution: Install Makrolon over the glass, (Case Study 4). New deadbolt and strike plate. New hinge screws.

OUR OLD FRONT DOOR

Glass panes (especially this many) weaken a door's structure. Glass can be broken, allowing access to the inside to unlock deadbolt(s) and the handle lock.

Old deadbolt throws a mere 1/2-inch. Way too short!

Doors are hollowed out to accomadate latches and deadbolts, weakening the area, especially in old wood doors.

Old pine door is brittle with age; panels are thin.

Three hinge sets attach to the jamb with 1-1/4" screws. Way too short! Screws barely penetrated the frame.

Our Solution: Replace with new door, (Case Study 3).

ZONE 2 CATALOG: DOORS

With newly acquired door knowledge, I could see that of our three exterior doors, the front one needed to be replaced. The other two needed serious reinforcement. Off I went to visit door stores, locksmiths, and online catalogs to learn my options.

The Zone 2 Door Catalog suggest ideas and products to increase door security. Two themes run through this catalog:

REINFORCE

Deadbolts. You must have at least one. If your deadbolt doesn't throw a full inch, replace it with one that does. Buy the best lock(s) you can afford.

Cylinders. If your deadbolt throws a full inch, but over the years you've given out copies of the key to your kids, dog sitters, and cleaning people–install a new cylinder. It's called re-keying. Read on.

Steel Reinforcements. Reinforce weakened areas along jamb and hinge areas.

Barred Door. For the money, it's the best interior door security for when you're inside your home.

Protecting Glass in Doors. Prevent someone from breaking the glass in or near your door and undoing your deadbolts.

REPLACE

New Doors. When you consider all the options, you may conclude that you need a new door. Wood, fiberglass, and steel are covered (See Appendix F).

I have a bias for physical security solutions not based on cell phones, radio frequencies, or bluetooth. For those loving the broadband life, Despair Not! Security solutions that you can install and monitor via the big grid, can be found on my website, calamitymay.com

DOOR LOCKS

Deadbolt Locks ANSI Grade 1

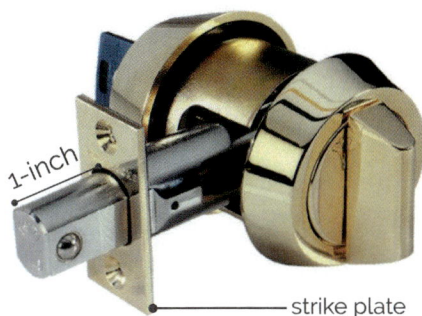

1-inch

strike plate

Look for the ANSI grade on the packaging. It's there. People working in building supply stores (local and national chains) are often unaware of the ANSI grade. Look for yourself!

The best exterior residential deadbolts are rated ANSI Grade 1. Their quality lies in tamper-proof design, superior locking mechanisms and materials, and good to great resistance to lock picking and bumping. The really good ones, like Mul-T-Lock have 10 pins in their tumblers and are highly resistant to jimmying, bumping, and picking. Schlage's ANSI Grade 1 has five pins. You pay more for Mul-T-Lock and to get key copies made you have to go to the locksmith who sold it to you with your identification! ANSI Grade 1 locks cost a little more and you have to look harder to find them. Locksmiths tend to carry better quality locks than big box stores. Shop around.

ANSI Grade 2

ANSI Grade 2 locks offer good to moderate resistance to lock tampering, they cost only a small amount less than ANSI Grade 1.

New Cylinder in an Existing Deadbolt

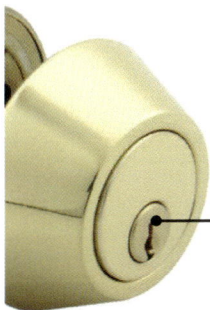

Replace the cylinder. Keep the deadbolt.

To replace the existing cylinder in your deadbolt (the part that receives the key), is called "re-keying". It's an efficient and cost effective upgrade. Bonus: if you re-key several doors at the same time, one key can be made to open all of them. Well chosen cylinders make a great security upgrade.

Cylinders can be sophisticated. All leading manufacturers make them. Here's a short list of Mul-T-Lock's features for residential cylinders:

- The cylinder inserts into your existing deadbolt, even if it's from another manufacturer.
- Materials include hardened steel and patented technology.
- Exceeds industry standards with 10 tumbler pins and two interlocking ball bearings.
- Only authorized Mul-T-Lock dealers can create Mul-T-Lock keys, even then you have to show up with your official Key Order Card.
- Burglars can't unlock this door from inside. (But make sure you can easily in case of fire!)

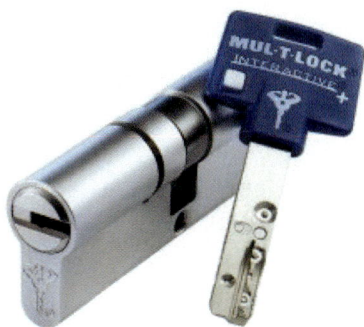

ZONE 2 – PREVENTION CATALOG: DOORS

DOOR LOCKS

Multi-Point Locking System

This is not a new idea, multiple point locks (MPL) have been used in European doors for a very long time. Though configurations vary, they are what they sound like. An MPL can have 3 deadbolts on the side of a door, throw a bolt up into the lintel, and one down into the door sill (floor). A twist of the handle extends all bolts into place.

These are highly engineered door systems. MPLs come in new doors, or can be retrofitted into existing ones. Either way, they're expensive. So is installation, as everything has to hit its mark. If your needs warrant the cost, look into them.

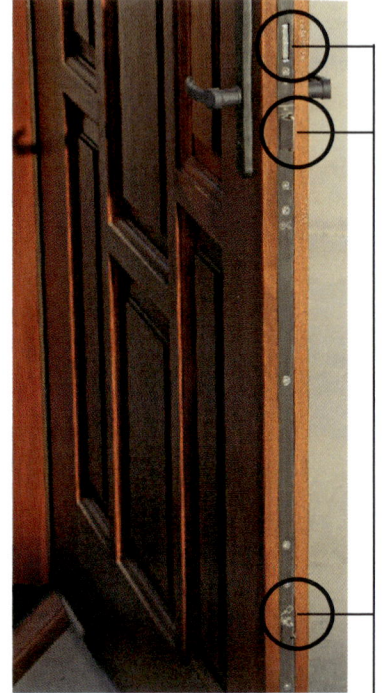

locks

DOOR & FRAME

EZ Armor's Max Combo Set: Steel Reinforcement

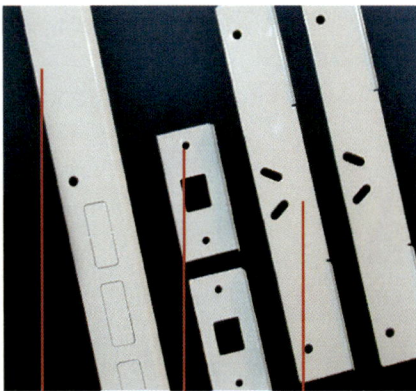

Edge of door cover protects deadbolt and handle area

Strike plate covers

Hinge cover (for 2 hinges)

The 2 weak zones of a door (Diagram, page 64) can be reinforced with steel that's barely noticeable. EZ Armor uses heavy, 12-gauge steel to armor these weak points. There are 2 reinforcement options:

1) Easy install / Good protection. The steel jamb covers over screws in the existing wood jamb. Great for damaged jambs if you're on a budget. The package, left, is $79 and installs in less than an hour.

2) Expert install / Great protection. A carpenter is needed to remove the interior door casing and install the steel over over the whole jamb.

The interior wall around the door will need to be refinished). This is real reinforcement. And more costly.
Both options screw into a door's structural frame.

EZ Armor sells individual parts and packages, like the one shown (bronze). Their steel reinforcements come in 3 finishes, and can be painted.

Their Lifetime Warranty against kick-ins demonstrates how thoroughly EZ Armor products have been tested. Read about it on their informative and instructive website, armorconcepts.com.

DOORS

Katy Bar®

An attractive, hurricane-rated door bar can be purchased from www.katybar.com. Their door bars are designed to be strong, yet easily removed in case of an emergency exit. Visit their Informative web site.

Barring doors works well on ground-level doors that a bad guy considers an easy target. (Dimly lit doors in the back of a house for instance, or, basement doors). I like them for exterior doors. Barring a door inside a 'safe room' in which one may retreat with children in case of home invasion, isn't a bad idea.

36" Katy Bar, MSRP. $225

DOORS

Door Movement Sensor

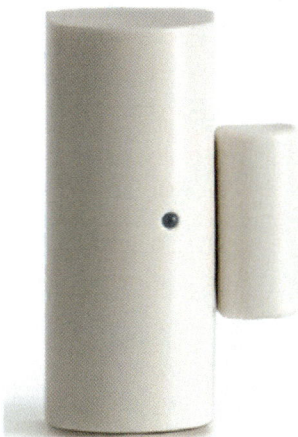

These paired sensors are components of home security systems sold by Simplisafe. They detect door movement greater than 2 inches. The small one attaches to the door, the big one, to the door frame (casing).

When a door opens, the magnet is disrupted, and a signal is sent to the receiver (in the house), and a chime rings. (Receiver on page 80). The sensors will still sound if the home security system is turned off.

These sensors work the same way on doors and windows. They come backed with adhesive, and batteries installed,

We've use them on the first floor for 4 years, they work great. You can buy only the sensors and receiver, but check out all SimpliSafe's options.

One pair, MSRP $14.99.

Receiver sold separately.

CASE STUDY 2:
OUR BARRED DOOR

Rustic barred door in our basement. Because the bar fits snugly, we keep a mallet next to it at all times to open in case of fire or emergency.

Why a barred door? It's a great way to Deny Access! Here's how this one works: Steel brackets screw into the structural door frame. A 2" x 4" piece of lumber spans the width of the door and fits into the brackets. That's it. They can be easily customized.
Two that we have:

Basement / Side Door: *Off the shelf, super simple.*
See photo above. A set of steel brackets are screwed into the structural frame with 3" screws. A 2" x 4" piece of lumber spans the brackets. These 3 pieces and 6 screws from a building supply store cost $18. I paid my nephew $20 to install it. (I do stuff like this when Joe's not home!)

Living Room / Main Entry. *Custom fabricated.*
Joe designed 2 heavy gauge steel brackets and had them fabricated. He painted them white like the trim. The brackets were designed to be removed when we entertain; in their place, decorative covers go over the raised screw heads. The painted 2" x 4" bar goes into a closet. Our medieval technology goes unnoticed.

I sleep better for our two barred doors. When Joe goes fishing, if someone tries to get in, they'll make enough noise to be greeted with a game changer.

DIY: Rustic Barred Door Recipe

Ingredients:
(2) galvanized U-brackets
(4) 3" long screw, #8 stainless
(1) 2"x4" board long enough to span both brackets.

• First establish a level line below the lock or deadbolt. Mark where the brackets will go. Place below any window in the door.
• With an electric drill, screw each bracket through the casing into the structural door frame with 3" screws.
• Slide the wooden board into place.
• To color match: prime and paint before installation.

A barred door must be easily opened in case of an emergency exit or fire.

DIY level: *Basic*

CASE STUDY 3:
NEW DOOR PURCHASE

Replacing our old front door was, for me, an unhappy but necessary upgrade. I loved its weathered charm, many beveled glass panes, and brass hardware. But seeing it through the lens of these security lessons, it was fraught with weakness. Because the old door jamb was still in good shape and securely fastened to the frame, we decided to replace only the door. I released my sentimental attachment and began the quest for a new front door that was strong, attractive and within our budget.

At my first stop, Allegheny Millwork, I was quickly burdened by overchoice! Wood is attractive. Fiberglass looked surprisingly good. Fortunately Joe, an architect, found one that fit our house, budget and security needs. When it arrived, he hired Adam, a seasoned carpenter, who can hang doors in his sleep (while smoking). When Adam started, he pointed out that our lintel was crooked (*What?* Never saw that! And I painted it twice!). He asked if we wanted the new door to be modified to fit this imperfection. Lesson learned: only wood doors can be planed or modified to fit off-kilter frames in older homes.

Note on Materials

If you're considering a new door for your home, there are many, many considerations to take into account. Each type of door material, wood, fiberglass, or steel, has pros and cons. Read about them in Appendix F.

Don't forget to outfit your new door with an ANSI Grade 1 deadbolt and a wireless floor mat (page 58).

DIY level: *Advanced*

CASE STUDY 4:
PROTECTING GLASS

A large expanse of glass in a door is great for light and views, but it poses two security problems:

1) it can be broken, the lock undone, then walked through by the uninvited; and,
2) a large glass expanse of glass weakens a door's structural integrity.

Pre-Home Security Research

After we bought our house, we transformed a solid wood door on the back of our house by installing a large pane of glass to see out to the river valley beyond. Joe removed a 50" x 20" area of the wood panel and installed tempered, insulated glass of the same size. The resulting light and views were great.

Post-Home Security Research

Two years into the research of this book, we were about to leave on holiday. We were pretty well along on our security upgrades, except for this rear door. We looked at it anew, and knew something had to be done to secure the glass.

Polycarbonate (Plastic glass)

Joe decided to add a layer of breakage protection over the glass. After researching materials and availability, we chose a local supplier, Laird Plastics, that sells Makrolon®, a crystal clear, hard-to-break polycarbonate. (Think: plastic glass). We chose 1/4" thickness and had it custom cut 3" wider than the 50"x 20" tempered glass. The extra 1.5" on all sides left enough room to attach it to the wood door's stiles and rails.

He chose mating bolts to secure it. At 2.5" long, they extend through the door and the Makrolon, and leave a 3/4" space between the plastic and the glass. (The 3/4" space allows room for the plastic to flex, under force of a blow, without touching the glass). Mating bolts have smooth exterior heads that can't be removed from the outside and are large enough to take an impact.

For fun, prior to install—with the protective paper still on—we beat the Makrolon with a hammer. To our surprise (and relief!) it barely flinched. It's installed, and the glass underneath, secured. We went on vacation with one less vulnerable area in our house. **Access denied!**

Plastic glass (such as Makrolon®, Plexiglas®, or Lexan®) creates see-thru barriers that demand special tools, time, trouble and noise to overcome. Great deterrents for vulnerable windows in Zone 2. Can be applied to a door's glass side panels as well.

We installed 3 more at home, and on rental property.

DIY level: *Advanced*

ZONE 2 CATALOG: WINDOWS

Windows and categories of glass is an even broader topic than doors, historically and technically. But let's keep it simple. This part of the Zone 2 catalog focuses on preventing illegal entry through popular double-hung windows.

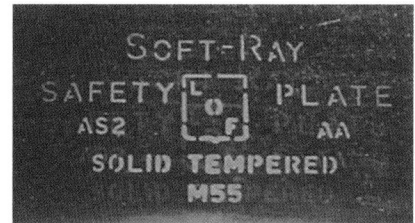

Since 1976, residential windows have been made of tempered glass. It's much stronger and breaks much more safely than residential glass prior to that year. All tempered glass bears a 'mark' in one of the corners, similar to that above. Look for the word 'Tempered'. If you have windows or sliding glass doors not made of tempered glass, know that the glass will break more easily (than tempered), and is likely to break into many very sharp, dagger-like points. (Never lean on untempered glass in historic buildings!)

The Zone 2 Window catalog will help you think about viable options you can use. Three themes underlie the products in this part of the catalog:

SOUND

Audio alerts. Hear a chime sound when your windows move up or down. Know when one's been broken. (Home security system involved).

LOCK

Lock your windows shut; or, lock them into a 4-inch 'open' position.

SECURITY SYSTEM

I talk about the DIY home security system that we use that's made and sold online by SimpliSafe. It works well, it's affordable, and straightforward to install.

NOTE: There are many professional home security systems on the market. If you go that route: research police response rates to home alarms where you live. Re-visit Myths 3 and 4 (you still need to Have a Plan). And, if you buy a system, use it consistently. Start by turning it on.

WINDOW SASHES & HOME SECURITY

Entry Sensor

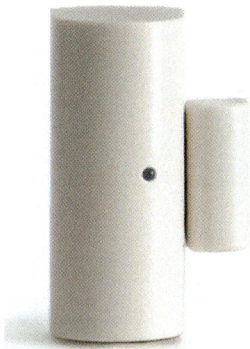

These paired sensors detect window movement greater than 2 inches. The small one attaches to the window sash, the big one, to the window frame.

If window movement breaks the magnetic attraction, the sensor sends a signal to the receiver in the house, and a chime rings. (Receiver shown below). The sensors will still sound if the home security system is turned off. These sensors work on doors too.

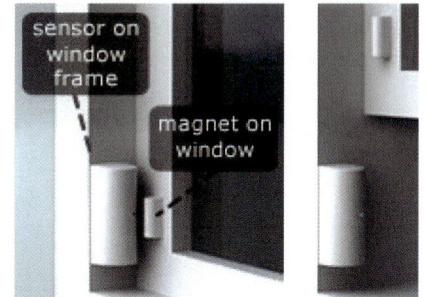

sensor on window frame

magnet on window

Both sensors come backed with adhesive, and batteries installed,

(Adhesive lost its grip after 3 years, easy replacement).

One pair, MSRP $14.99.

Glass Break Sensor

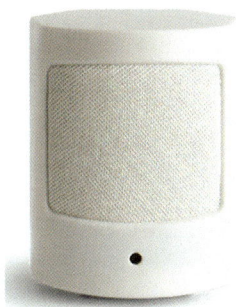

Breaking glass has a complex sound 'signature'. It took the home security industry years to develop residential sensors that isolate the sound of breaking glass from shattering plates or chirping birds.

SimpliSafe figured it out too. Each sensor covers 20-feet in all directions.

Each, MSRP $34.99

Receiver

SimpliSafe's two sensors on this page relay information to this 11-inch high receiver that works for the whole house. We use the entry sensors on all our exterior doors and first-floor windows. Simplisafe sells small and large home security packages that you customize with individual components. It's easy to add pieces later as you need them.

SimpliSafe sells directly to consumers through the web. They have great prices; good products; and an informative, online user forum.

Installing their system is easy, definitely DIY. Joe installed our whole house system in an hour. Professional monitoring, 24/7/365 is optional and available. Price varies based on configuration. Visit www.simplisafe.com.

WINDOW SASHES

Window Pin Lock

A steel pin prevents the lifting or sliding of a window. It can also securely lock a window into an 'open' or ventialting position. (A 4" maximum opening is recommended). Be careful to drill a hole outside the edge of the pane of glass.

Stainless steel window pin locks are strong and cheap. Primeline. net has an extensive selection of window-locking solutions. (See their Window Security section). Check Lowe's or Home Depot's web sites as well.

If you're good at not losing keys. A keyed window latch will not let anyone open the latch without the key. Even bad guys with advanced tools may be deterred.

DIY: in double hung windows drill the same hole through the upper and lower sash. Do it when it's closed, and lift the window and drill a new shared hole for the open, locked position. Insert pin to lock it open. Do not drill through to the outside of the window frame!

WHAT WE DID AT OUR HOUSE
A 3 YEAR LATER EVALUATION

The following page lists security upgrades we made to our small home on an 1/8-acre lot in the city. Over 3 years we added what we could afford. Our integrated approach to prevent break ins is comprised of detection in Zone 1; fortified doors and windows in Zone 2; breach detection in Zone 3. You'll find the actions we took, costs, and ease of install. I added my satisfaction (Happy) ratings and comments. Our biggest expense: replacing old, drafty, non-tempered windows with 10 new, high-quality windows is not shown.

NOTES

Much of this stuff is easy to install if you're a diligent novice willing to use a drill. Joe's a proficient DIY guy, handy with advanced tools and likes to undertake projects.

3 levels of DIY (for home security installation) defined.

1	You open product packaging, read instructions, install. No tools necessary. It can also mean developing a new habit
2	Requires light to moderate use of power tools, namely, an electric drill, and using it in forward and reverse.
3	Advanced carpentry skills; experience with power tools and thoughtful execution of projects. A close-to-expert to expert level. (A cake walk for good carpenters and cabinet makers!)
Pro	Experienced, professional carpenter.

What Would I Do Differently?

I'd spend money on an electrician to hard wire motion sensor lights in Zones 1 & 2, instead of relying on cheap, short-lived, solar-powered units.

What Was the Most Fun?

Learning how to use an electric drill to take out door hinge screws and measure them!

A BALANCED APPROACH

I'm not into spending a lot of time, money, and energy on home security. Joe and I think about it, we choose tools and develop new habits in order to create a positive environment with less subconscious stress.

If a tool doesn't suit us, we try something else, make adjustments, and get on with life...it's all good.

	Item/Action	Material Cost ($)	Ease install 1-3	Happy 0-5	Comments
ZONE 1					
Outside	Solar-powered Motion Sensor Lights	160	1	2	Cheap solar units lasted 4 months
	Remove Tools in Yard that could cause harm	0	1	5	Still mindful and bringing tools in!
ZONE 2					
DOORS					
Old Front Door	*Interim Solution:*				
	Barred Door w/ custom brackets	60	3	5	Physical barrier to entry.
	3/8" thick Makrolon over 24" x 30" door glass	80	3	5	Still 95% transparent; very secure.
	SimpliSafe DIY home security system (door & window sensors)	see below	1	5	Still works great after 2 years, Velcro on sensors gave way. Easily replaced.
New Front Door	Wood Cor-Stave Door	1300	pro	5	Attractive. 3 windows across the top
	ANSI Grade 1 deadbolt & handle	85	3	5	So far, so good!
	Professional install of door & lock	350	3	5	Experienced carpenter hired.
	(9) 3" screws into hinge plates	14	2	5	Easy to do with electric screwdriver.
	(2) 3" long screws into strike plate	3	2		Easy to do with electric screwdriver.
	Door motion sensor (reinstalled)	see below	1	5	Still flawless. Good to hear door opening.
	Barred Door w/ custom brackets	reused	2	5	Low flex, holds strong.
Basement Side Door	Barred Door w/ basic brackets	20	2	5	Love it! Nephew installed. Mallet kept nearby to unloosen in emergency.
	3/8" thick Makrolon over window	60	3	5	Unbreakable. Lets light into basement
	New deadbolt, ANSI Grade 1	40	3	5	Works perfectly.
	(4) 3"-long screws into hinge plates	6	2	5	Easy to do with electric drill.
	(2) 3" long screws into strike plate	3	2	5	Easy to do with electric drill.
	Motion sensor in basement stairwell	see below	1	3	Removed. Bad placement. Dogs set it off.
	Door motion sensor	see below	1	5	Consistently chimes when door opens
Rear Kitchen Door	Makrolon over 21" x 50" window	95	3	4	Very secure. Lost 3% transparceny. Hard to access to clean, but looks pretty good.
	New deadbolt, ANSI Grade 1	40	3	5	Works perfectly.
	(4) 3" long screws in each hinge	6	2	5	Easy to do with electric screwdriver.
	(2) 3" long screws into strike plate	3	2	5	Easy to do with electric screwdriver.
	Door motion sensor	see below	1	5	Consistently chimes when door opens
WINDOWS					
Rear Window	Joe fabricated Old World wood shutter that latches from inside	75	3		Sturdy. Hardware attachment to brick could be better.
	Window motion sensor	see below	1	5	Works consistently.
First-Floor	5 Window motion sensors	see below	2	5	Works consistently.
	Curtains on windows facing street	120	1	5	1st upgrade. Blocks visual access into house.
	Glass Break Sensor	35	1	5	No accidental alarm.
2nd-Floor	4 Window motion sensors	see below	1	5	Works consistently.
	Window Pins, on all 1st-floor windows	20	2	5	Drill away from edge of tempered glass.
Basement	4 custom awning windows, tempered glass. Includes install	800	pro	5	Excellent quality; work great.
ZONE 3 Alert System					
Home Security System	Motion sensors-all exterior doors & windows; 1 interior motion sensor; 2 panic buttons; exterior siren; 2 keychain remotes.	320	easy	5	We use only for motion sensor notification. Not monitored. Works great. Easy install. Keychain remotes flimsy.
Total		**3695**			

Still dabbling in denial thinking?

But I live in a nice neighborhood

Revisit the 5 Myths

I've never heard of trouble where I live

They're called taboo topics for a reason

I can't afford anything right now

Locking doors & windows: free. Barring a door: $20

I have no idea how to install anything

Do I hear self-limiting beliefs? Ask for help. Watch YouTube. Take a DIY class at Home Depot

I'll never use it

Try things that make sense for how you operate. If it doesn't work for you, return it

My kids or husband will never get on board

Speak to them with facts, certainty, and self assurance. Let them watch you wield an electric drill!

REVIEW & ADVICE

- Learn what threats are in your neighborhood. Get objective information.

- Read the annual police report for where you live.

- Go to the local police station and ask types of crime and frequency.

- Keep your ear to the ground at local meetings, read the paper.

- Talk with your neighbors…what are they seeing or witnessing?

- Develop prevention plans in response to crimes in your neighborhood. Tailor them to work for who you are and how you live.

- Start small. Familiarize yourself with each security component until you're comfortable with it. Then, add the next one. Make incremental improvements as you can afford the time, money, or energy. (Many are free or cheap!)

- Be willing to try new things. Trust that you'll know which products are right for you.

- Finally, doing nothing is a choice. But I don't recommend it.

For women who live alone, or who never thought to look, or don't know how to assess the security of one's home: get up and go look at your doors and windows. See with fresh eyes: jiggle the door latches, remove a screw from a hinge plate and measure it….. If it's less than 3", you know what to do.

Create an integrated security system that works for you. Phase it in. Fortify your castle—You **Are** the Keep.

WHY IS THIS SO IMPORTANT? Why do I harp on this topic? BECAUSE,
YOU WANT PREVENTION
MORE THAN YOU WANT SELF DEFENSE

> "Above all,
> be the heroine
> of your life,
> not the victim."
>
> *– Nora Ephron*

1 You work from home and just nailed a deadline. You take a break and go outside to stretch. Back inside, you lock the sliding glass door and decide to throw in laundry. You're down in the basement when you hear footsteps overhead. But nobody's supposed to be there! You go quiet, and hear rummaging...something broke! A stranger swore and stopped to hear if anyone would respond. Some brave part of you leads you to that side basement door up a half flight of steps...The hard-to-open door gives way in a millisecond and you're out and running down the street to Ann and Roger's house, hoping they're home...

2 It's 3am, you're sleeping, so is Fido on your bed. You didn't hear the window slide open downstairs—neither did the snoring dog. Not until you heard "a bump" downstairs do you rustle. Very quickly, you're hyper awake. The dog picks up your cue, and starts to bark but doesn't go down. No matter. You've thought about this. You hit the panic button on the car keys that you keep next to your bed. Your cell phone's there too, you call 9-1-1, leave the phone open, so the dispatcher can hear you loudly and assertively tell the intruder to leave, that you've called the police, that you're armed and prepared to shoot. (In Pennsylvania, you can do this). You do not go downstairs. You have a plan and backup plans. One of which is a strategic location in the bedroom for defensive shooting...if it's comes to that....

PART III

SELF-DEFENSE

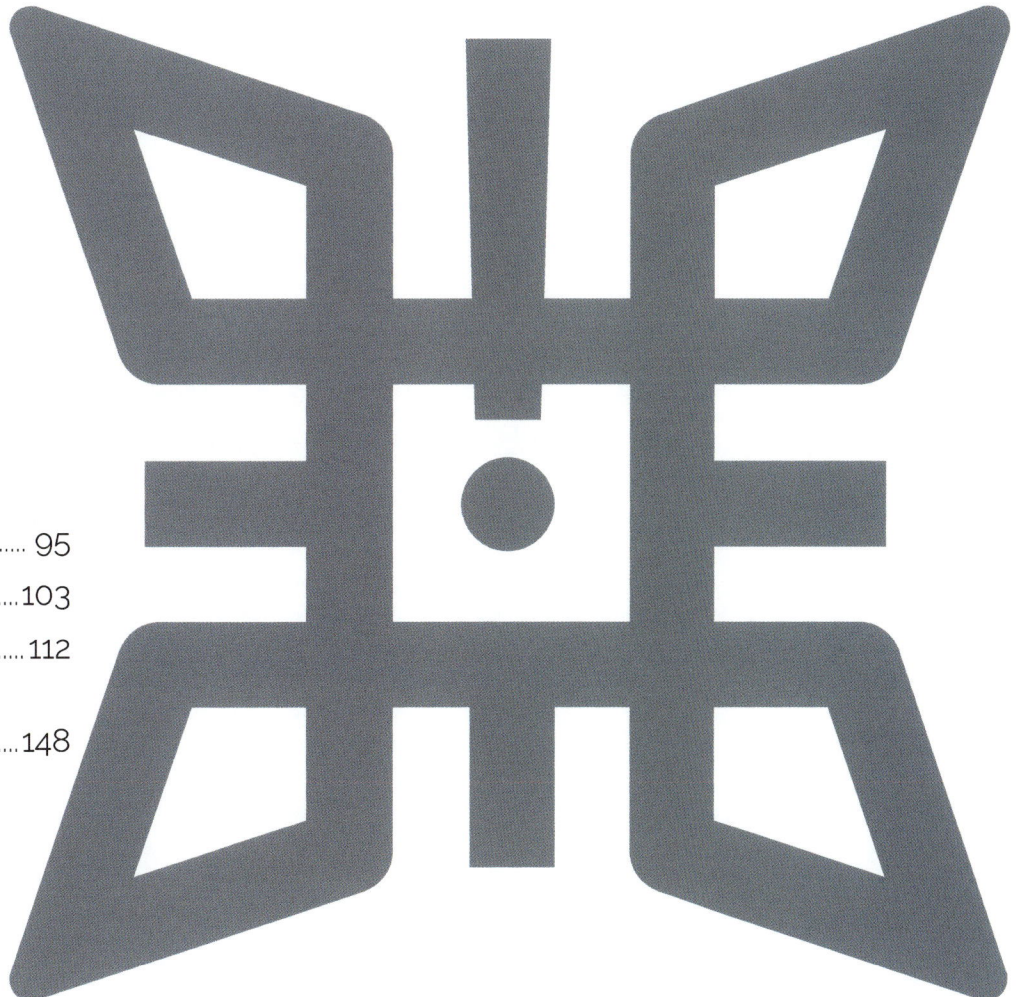

In Zone 1, we looked at ways to eliminate surprise.

In Zone 2, we covered how to deter access.

In Zone 3, we consider home & self defense. Have a plan when someone breaches your castle walls. What will your response be? What are you willing and able to do?

Smart, integrated security measures go a long way to deter predators. Despite your best efforts, however, a bad guy with enough time, tools, or determination may work his way into your home. You could be the only one able or willing to stop him. The moment you hear that bump in the night is not the time to design your defense plan. The time is now, Home Defender.

In this section, I shine a spotlight on your powerful survival instincts. I take you through a Willingness Exercise to develop a life-saving mindset. Numerous self defense techniques are illustrated: pepper spray, improvised weapons, unarmed self defense, and firearms.

Develop your own integrated self-defense plan and choose tools that work best for you.

DISCLAIMER 2: The following pages express the personal thoughts of the author only. They are not offered as suggestions or advice to any other person. Every event, person, home, circumstance is different. What you are willing to do, choose to do, and act upon or not, is solely your reponsibility, and in no way reflects the material presented here. See Disclaimer 1, Frontispiece.

ZONE 3: INSIDE YOUR HOME

HAVE A PLAN

Home & self defense if your castle walls are breached.

To effectively defend your home, family, or self, you need the right kind of mindset and willingness to act. You need a plan, backup plans and the fortitude to see them through. The good news is, your ancestors left you a legacy of survival gear. It's inside you, waiting for your attention.

INTRODUCE YOURSELF TO PRIMAL YOU

Built for fight or flight—like your foremothers and forefathers in your ancient family tree—you're innately well equipped to handle the rare stresses of enemies and catastrophes. Your body, your whole being evolved with abilities to detect and respond swiftly to hair-raising situations. These responses come from your gut, heart, and brain–a legacy of survival instincts inherited from your ancestors.

Embedded in the fabric of your being is an ancient, breathing human animal. Its function: to make sure you stick around. Some call it survival instinct. I call it Primal You. You're hardwired for fight or flight. Primal You probably saved your hide or a friend's several times already, reflexively pulling her back as she stepped off the curb at the wrong time. Primal You sleeps with one eye open and is always On. That's the big gift from your ancestors.

Though naturally sturdy in nature, Primal You is susceptible to neglect (from a sedentary lifestyle); soft decay (from addictions to bad food, sugar, alcohol, and pharmaceuticals); and numbness (from mindless TV, 24-hour news, and relentless social media)—indulgences that diminish your primal connection to the natural world. Calamity May recommends that you take this moment to recognize Primal You. Say hi. Cultivate a relationship. Go for a walk in the woods. Take good care of Primal You.

HOW DOES ONE CULTIVATE PRIMAL YOU?

EAT WELL Skip futile diets, go for enduring lifestyle changes. Primal You is sustained on real foods that go back tens of thousands of years. There is no diet pop, fake butter, corn syrup, genetically modified organisms or purple pills in the Primal You diet. We're talking hormone-free meats, healthy fats, fresh vegetables, fruits, seeds and nuts. Stuff that farmers raise, hunters shoot, and gardeners grow. If you're overweight, constipated, have brain fog and no energy; if your joints hurt; if your skin is dull and your hair listless: rethink your diet! I did.

I love the many forms of bread (and pizza) as much as the next person, but I stopped eating it. My health has regenerated immensely. The above symptoms and conditions are gone, I dropped 22 pounds and have kept it off. Even my gray hair diminished! To learn how and why you want to eat like your ancestors for optimal health, I suggest you read *Primal Body Primal Mind* by Nora Gedgaudas, CNS, NTP, BCHN. Or go for a hard core re-set as outlined by Dr. Jack Kruse in his blog and book, *Epi-Paleo Rx*. Both authors write with passion from personal experience and deep research.

GET STRONG, DEVELOP CORE STRENGTH During a hunt, your ancestors walked, stalked, ran, dug, clubbed, speared, and dragged heavy burdens back to homebase, their bodies exhausted. With food in store, they relaxed and recovered.

By practicing exertion and recovery, (one form of which is High Intensity Training (HIT) or Super Slow Strength Training) we can achieve remarkable fitness and health. My warrior mentor showed me how and turned me onto Dr. Doug McGuff's book, *Body by Science.* It's extremely useful for learning the how-tos and benefits of a super slow workout and the importance of allowing 7 days for recovery (depending on your current condition). Here's what I really like about HIT:

1) I do it at home with my own body weight for 15 minutes, once a week. 2) The protocols are free. 3) They work. (I got a 2-inch waistline reduction after five sessions and all my joints feel better for it). 4) It applies to all ages, and greatly benefits seniors. Rejuvenate Primal You!

(Note: McGuff uses gym equipment. My mentor, an expert in biomechanics and trainer of elite forces showed me four basic exercises using only my body weight for resistance: squats, sit-ups, pull-ups, and push-ups. Super slow style. See how I do it at calamitymay.com).

GET STARTED WITH PRIMAL YOU

Distracted?

Out of breath?

Weak ankles?

Visit my fitness blogs at calamitymay.com

BECOME MINDFUL Your survival instincts live neither in the past nor the future. They function in the present. This very moment. Part of cultivating Primal You is practicing mindful Awareness. For example: five times a day, do one thing with thoughtful intent. Sit down to a meal, chew your food slowly and focus on the flavors. Drive without a cell phone. Walk outside without earbuds. Look up at the sky, see the quality of the light. Where's north? In which direction are the clouds moving? Are pigeons, seagulls, or hawks in the air above you? Focusing on the natural, physical environment that you're a part of, sharpens all your senses.

PRIMAL YOU PLAYS OFFENSE & DEFENSE If you're wondering how to wrap your head around being a Home Defender, consider what is precious to you: Your life? Your children? Your old Mum who lives with you? (Primal You says "You bet!") Your life's savings under the mattress? ("Heck Yeah!") Use your heart and brain to consider who/what you're willing to defend. What will you give yourself permission to do when you're home and someone breaches your castle walls? Will you fend off an intruder who enters with intent to rob, assault, rape, or torture you for your money and jewels? (It happens). This is no time to be polite! Turn the tables: with illegal entry, an intruder just endangered themselves!

Whether you wield a garden stake, a baseball bat, a frying pan, or a firearm, you just became engaged in a personal battle to defend your family and your life. It's okay to be afraid! It's not okay to freeze with fear. Go on the Offensive until they flee, or until they're stopped. There's so much you can do!

WILLINGNESS Most of us live in a day-to-day world so buffered from survival that we don't even know how much we've forgotten. Vital knowledge and hands-on skills were passed from generation to generation, handed down in scouting groups, fathers' workshops, and grandmothers' kitchens.

The mindset necessary for defending oneself from intruders, calls upon a life-affirming agreement to self and tribe. It also calls upon physical skill, cleverness, and the appropriate use of tools.

When I was eight years old, daydreaming up in a tree, I just knew that I'd bravely protect my Mom, Dad, older brothers, or anyone in my family from attack. As a little kid, I had the heart of a lioness. As an adult, nearly five decades later, I still lacked sufficient mindset and skills to stop a bad guy from doing harm. I now know how. This part of the book shares how I'm turning that daydream into reality. And you can too.

If you think you don't have what it takes to prevail in life and death situations – think again.

Call upon your foremothers and forefathers for support. Your mother's mother 200 generations ago may have kicked a real Neandrethal in his family jewels, and then some!

Here's to her.

Here's to all your ancestors who made it. You're here because of them.

EUREKA!

One of my mentors was a Major in the U.S. Marines Special Operations Command. He trained elite forces and is an immensely qualified and seasoned warrior. Usually once a month, I sit with him in a coffee shop and listen to brief glimpses into his 26 years embroiled in war, covert operations, civil unrest, training, and survival. One day I brought in books on self-defense laws. He dismissed them as 'minutiae' and went on to describe to me how to get rid of evidence. It was graphic and disturbing. "I'm not willing to do that!", I blurted out. Something struck. I said goodbye and left, got in my car and sat there. My stomach churned while I stared at sycamore trees. Then it surfaced: *What am I willing to do?* It was a watershed moment, when the Queen of Denial put her toe in the stream of a warrior's mindset.

In the context of self defense, willingness means giving yourself permission to mentally and physically prepare to take decisive action in a necessary instant. It's a mindset that may be instinctual, but is critically reinforced when you think about those actions and practice them ahead of time. Exploring what you are willing to do is a no bullshit exercise. I'd say it's the crux of the biscuit.

What if the fight does come to you? It has many guises: the jealous ex-boyfriend, the bully, the fawning schoolteacher, the thug in your subway car. What if they lock onto your best friend? If the situation doesn't de-escalate, or you can't walk or run from it, consider the threat imminent—a lunge away. You're now in a personal battle. To decisively fight, prevail, and survive, it behooves you to have thought about this well before a dark moment casts its shadow on you. You have to be willing to stand up to a menacing person and do what it takes to stop him/her. Most of us are so buffered from real survival that we never learned how, or have forgotten what that means. Here's the refresher course.

KNOW WHAT YOU'RE WILLING TO DO AND WORK ON THAT

Early in my research for this book, gunsmith and writer, Ed Siemon, lent me Jan Jones' book, *Self Defense Requires No Apologies*. I was bowled over by her plain talk and woman-to-woman tone. She wrote it in 1985 yet every word is timeless. Example: "If you follow the suggestions in this book, you will be luckier than most women. You will profit from the experience of those who have been assaulted without having to experience it yourself. You will possess the skills, and hopefully, the willingness to defend your life if necessary. The paradox is that if you have this ability, you will probably never be forced to use it". Her book is highly readable and recommended.

In the course of my journey from Queen of Denial to Home Defender, I've thought hard about prevention, preparation, and defense. I made up scenarios (from easy to heart pounding), worked through possible reactions, and wrote them down. As the list grew, so did my certitude. I call it the Willingness Exercise, and it continues to be my touchstone for what I give myself permission to do. If you're looking for a starting point, to cultivate a home defender mindset, see page 96-97. The exercise looks orderly now, but it was rough going at first. Figuring out what works for you calls upon your gut, heart, and brain for answers.

There's one more layer that you need to be aware of: the self defense laws of your state. They vary widely state by state. (This is lightly touched upon at the end of Part 3, page 154). Why does this matter? Because what you're allowed to do according to your state's laws, may temper what you're willing to do.

Example: I'm not a big materialist, and I'm not going to shoot a young teenager who's half way over the fence with my laptop. I'd let him go for moral reasons. Here are two other points: in Pennsylvania, I could use force to retrieve my property, as in running after him and tackling him, but I can't use deadly force. Second, juries in all states will frown upon shooting an underage person (in possession of your property) who's running off. I've wrapped my head around Pennsylvania laws regarding home intrusion and self defense. If a determined intruder breaks into our house at night seeking pharmaceuticals in the upstairs bathroom—I have a different plan. That plan is grounded in the answers to these questions:

What am I willing to do?

What does my state allow me to do, legally?

To consider the former without knowing the latter can end very badly if defending your home and the lives in it means losing the legal battle and living even a few years in the hell of prison. In addition, depending on your state, circumstances, and defense attorney, you could lose the people/valuables/home that you were defending in the first place.

HERE'S MY SHORT LIST OF WHAT I'M WILLING TO DO. I WILL:

- I will install and use preventive measures to keep intruders and criminals out of my house.

- I will learn to counter threatening intrusions and violent acts toward myself and family.

- I will learn my state's laws about the 'reasonable' extent(s) to which I can exercise self defense, whether I agree with those laws or not.

A short list of What I **am not** willing to do:

- I won't be talked out of the need for prevention and self defense.

- I won't be unfit, unhealthy and/or unprepared to respond quickly.

- I won't relinquish responsibility for my personal safety.

Three years ago I didn't have this mindset—*believe me!* Since then I've morphed from a state of comfortable denial into a mindset that calls for personal responsibility and personal safety. I've taken classes in martial arts and firearms training. I've sought input from retired Green Berets, Rangers, Marines and active duty policemen. I've read well-regarded books by self-defense experts. I've sifted through a mountain of disheartening data to understand trends in crimes against people and property. I've been moved by the personal accounts of women I know who were raped by once intimate partners or strangers. It boils down to this: Be Aware, Follow Your Intuition, Avoid Danger, and Know How to Be A Lot of Trouble.

Work on what you're willing to do, then train your mind and train your body. Practice. Get confident. Work with your partner, kids, roommates, neighbors, and Aunt Favorite. Allow yourself to think differently. Visualize fighting for your life and emerging alive and whole. Let Primal You do what it needs to do—what you've trained to do—in the 4-7 seconds you have to get out intact, or stop the bad guy in your home.

Your personal safety is yours and yours alone to take care of. Never for one minute, expect someone else, your husband, son, dog, policeman or government to do it for you.

If you see where I'm going with this, remember:

YOU WANT PREVENTION

MORE THAN SELF DEFENSE.

WILLINGNESS EXERCISE

Note of Caution: This exercise can be gritty. At the outset, it affected me to the core. Fortunately, Primal Me eagerly showed up to assist. The Willingness Exercise asks, "What am I willing to do & How do I make it happen?" The worksheet I developed is filled with my responses. I strongly suggest you make up your own scenarios, or use mine to spur you on. For each reaction to a scenario, be specific in how you'll make it happen. For instance, "I don't want any bad guys breaking into my house." Therefore: "I'll install locks on all my first floor doors and windows and use them consistently." Be honest. Listen to your body for answers, it's smarter than you think.

Willingness Exercise

I AM WILLING TO:	HOW I'LL MAKE THAT HAPPEN:
Bear responsibility for my personal safety.	Stop assuming nothing bad will ever happen to me. Stop assuming my house isn't vulnerable to intrusion. Never assume that someone else, police, or government will "be there" to protect me from harm.
Do my best to prevent home break-ins.	Implement Zone 1 & 2 tools and techniques that make sense for our house and how we live; and, are relevant to threats in our area.
Develop and maintain core body strength.	Read: *Body by Science* by Dr. Doug McGuff. Practice H.I.T. once a week. Jump on my rebounder 3x/wk. Climb hills when walking the dogs. Take steps instead of an elevator.
Understand what it takes to reduce inflammation and joint pain, maintain a healthy cardiovascular system, and keep my brain in good shape.	Eat clean meats, fish, and fresh vegetables and fruit. Greatly reduce processed foods and sugar (including alcohol).
Cultivate a calm mind and grounded presence for daily life, and, greater mental stability for emergency situations. Recognize and reduce daily stress.	Meditate daily; practice Tai Chi daily; do yoga 3x/wk. Focus on breathing through my lower belly.
Eliminate potentially dangerous tools that could be used against us from outside our house.	Bring in Joe's hatchet from the woodpile; store sharp gardening tools inside.
Work together with my neighbors to increase neighborhood safelty.	Start a phone tree to alert each other of strange activity and behaviors on the streets around us. Contact my local police station's Community Officer to offer guidance in starting a neighborhood watch group. Invite neighbors to participate. Know which neighbors are prepared for self defense and those who are vulnerable and share plans to help. (Have their numbers on speed dial and vise versa).
Research self-defense laws in my state. Know what's legally defensible regarding self defense in my home. (*Tricky and time consuming in Pennsylvania!*)	Read the most recent editions of: *The Self Defense Laws of 50 States* by Vilos; and, *The Law of Self Defense* by Branca to understand interpretations. Stay current with state statutes online. Locate the best local criminal defense attorney. Introduce myself and keep business card handy. Join U.S. Law Shield, attend their free seminars.

I AM WILLING TO:

Learn what a "defensive position" is. Locate them in my home.

Learn specific actions I can take to defend myself and family members.

Understand that after a very grim encounter in my home, where an intruder is dead or wounded, that I could make incriminating statements that would be used against me.

Emerge from assault intact!

HOW I'LL MAKE THAT HAPPEN:

Learn how the layout of my home and property informs strategies for defense by identifying viable options for a safe room, retreat, concealment, or escape. If using weapons (improvised or firearms) know where and how to best to use them. Practice.

Unarmed - Take Krav Maga, Bulldog Jeet Kune Do, or close quarters combat, 2x/week for at least 2 years.
Hit my car key's panic button to scare off an intruder.
Stop a would-be rapist on top of me by gouging out his eyes.

Improvised - Identify kitchen 'weapons': cast iron skillet, scissors, hot coffee, knives.
Use the corner of my cell phone to crack an assailant's head.

Armed - Keep a shotgun next to my bed.
Use ammunition that can stop a bad guy but not penetrate walls.
Continue with gun safety and shooting courses taught by highly qualified instructors.
Get comfortable with cleaning my firearms.
Practice weekly, develop muscle memory.
Keep up my chops as a certified NRA instructor.

Have a really good defense attorney at the ready (one fluent in my state's self-defense laws).
Know what minimal and necessary information to convey to the 9-1-1 dispatcher and police when they arrive, and say no more.
Call my defense attorney immediately after calling 9-1-1.

Cultivate a warrior mindset.
Have a plan and multiple backup plans !
Visualize defensive tools in each room of my house.
Know my escape routes and how to get out, fast.
Practice.

CALAMITY MAY RECOMMENDS

I AM **NOT** WILLING TO:

- Be lazy and not use or maintain the Zone 1 & 2 measures we installed.
- Be caught in a potentially dangerous situation in my home without a plan.
- Lack mindset and training to defend myself
- Go down without a fight.
- Lack the stamina to run five blocks.
- Live in fear and paranoia.
- Be handicapped (broken foot? cut off finger?) and not have an appropriate weapon within reach.
- Follow bad advice. Example: drag a wounded intruder into the house if he falls outside of it. (It's called tampering with evidence, it's illegal).
- Let anyone, including myself, handle or use firearms after drinking alcohol, or taking any over-the-counter meds that impairs one's ability to drive or operate machinery.
- Leave my gun in plain sight when I'm not home, or while people are visiting.
- End up in jail because I inadvertently made self-incriminating statements to 9-1-1 or police.

"I thought it was a real, distinct possibility that I might be killed. I believed he would kill me if I resisted. But the other part was that I would try to kill him first because I guess that for me, at that time in my life, it would have been better to have died resisting rape than to have been raped. I decided I wasn't going to die. It seemed a waste to die on the floor in my apartment so I decided to fight."

– A woman who successfully resisted rape.[1]

UNARMED SELF DEFENSE

Our bodies haven't changed much over 40,000 years.[2] Vulnerable areas remain the same. Anyone's hands, feet, knees, elbows, or head can deliver forceful blows or inflict pain. Well-placed strikes and kicks to nerve centers and pressure points can bring big men down by smaller would-be victims.

Unarmed self defense has been around a very long time. Homo sapiens have evolved knowing how to defend from attack, and deliver strong counter offensives. Your greatest advantage: donning the warrior mindset to prevail.

Prior to that first wrist grab, or that hold from behind, two things can help you emerge alive and intact:

MINDSET:

The Pre-determined Will to Live and Fight; Primal You.

ACTION:

Surprise, Speed, and 100% Commitment.

Self defense is a normal, moral act.

– Jerry VanCook

How you put together strikes and blows in rapid sequence is why you sign up for classes in close combat, like: Jiu Jitsu, Krav Maga, Bulldog Jeet Kune Do, Systema Spetsnaz, or mixed martial arts. (Try several until you find one that works for you).

The self-defense techniques I'm about to describe are not new. Some, recorded since the Roman Empire, were taught to our fathers and grandfathers going into World Wars I & II, and are still taught today. The labeling of these enduring self-defense systems vary: practical defense, close combat, hand-to-hand combat, or street fighting. (*Off putting I know! Hang in there!*) They endure because they're effective, easy to learn and retain. These techniques aren't about fairness or sport. Nerve strikes, for instance, are not allowed in martial arts competitions because their effects can inflict long-lasting damage. Form doesn't matter here. There are no trophies when you're fighting for your life.

If you can avoid trouble, or walk away from it, do so. If attacked, proper mindset and moderate proficiency in self defense, can give you a good chance to prevail.

Some people go around armed at all times with a preferred weapon: a can of pepper spray, a cane, a gun, a knife, a metal pen or a bolt in their pocket. Most people, however, choose to not be armed, or dismiss the idea altogether. In regards to self defense, instructors, authors, and undercover cops will tell you this: weapons exist for a reason. A properly applied weapon can equalize the differences between a 100-pound girl and 200-pound bad guy, or a 70-year old man and a 32-year old speed freak. If the fight comes to you, and you have a weapon at hand: Use It! If you're caught without one, or choose not to have one, your situation just got even more difficult. So be it. Fight back. Use what you have.

YOUR BODY'S WEAPONS

Unbeknownst to you, dear reader, you're packing heat! Your hands, fingers, elbows, knees and feet can deliver real pain! If you know HOW and WHERE to strike, you can deter an assailant from further threat; leave a bad guy in writhing pain; or immobilize a home intruder—before the cops arrive. Know your body's weapons and strengthen them. Develop acumen through study, visualization, and practice.

> What we fear doing most is usually what we most need to do.
>
> – Ralph Waldo Emerson

Most thieves don't randomly enter a house with intent to harm the person(s) inside. Typically, they want a quick entry and exit with your stuff. However, police reports frequently show that things can end badly if someone's home. In a meth second, a doped-up criminal can change his mind from stealing stuff to inflicting harm. If an intruder is closing in on you fast at 20 feet or less, they'll overtake you in 2 seconds or less. The fight for your life is on.

With an accessible weapon in your hand, you can buy time. Hot coffee in the face; sand in the eyes. Reach for what's close: crochet needles, pepper spray, a cast-iron pan, a fast-drawn firearm. (You may get one shot off). If you're caught empty handed, get ready for a close encounter.

COUNTER-INTUITIVE I know it sounds counter-intuitive, but a close-in battle can give you opportunities to overcome an assailant's sheer strength in unexpected ways. Don't meet his strength with your lesser strength. Instead, use surprise to your advantage with quick, unexpected moves.

'Bulldog' Jeet Kune Do for instance, uses a defensive/offensive clenched double fist (page 111) that you slam into his Adam's Apple; follow with a chin jab and groin kick. It depends what body parts are open. A mighty kick to the kneecap may slow him down. If he's bent over grabbing his knee, deliver a blow to the base of his neck. If you're in a hold from behind, grab a finger and break it; stomp on his foot, crush small bones. Putting moves together within split seconds is what lets you prevail. Every situation's different. Whether you're tall or short, "Look for the openings." Speed, surprise, accuracy are your allies. Two years of quality martial arts training is a good minimum. Being able to deliver a few well-practiced moves based on your stature and abilities is a must.

BE A LOT OF TROUBLE If, from the beginning, you're a lot of trouble or show significant resistance, he may wise up and run out. If not, stay engaged in your personal battle until he is immobilized on the floor, or out the door. Because milliseconds matter, do what you must to end the encounter, or escape.

Two considerations: First, don't be faked out if the bad guy is down or pretending to not move. He may be waiting for you to turn your back and grab you or your weapon. Remain vigilant until help arrives. Second, a doped-up bad guy may not feel broken bones or deep wounds, and still come after you. Do nothing half-heartedly.

Coming up, a list of 10 body weapons and 15 targets. Figure out which ones may work for you based on your abilities and strengths. Use these moves only when you have to, like when a threatening person lays a hand on you, or tries to take you to another place. Don't use them on your drunken brother-in-law at a wedding.

Let's take some basic classes.

!

NEVER, EVER LET YOURSELF BE TAKEN TO ANOTHER PLACE

HISTORIC PERSPECTIVE

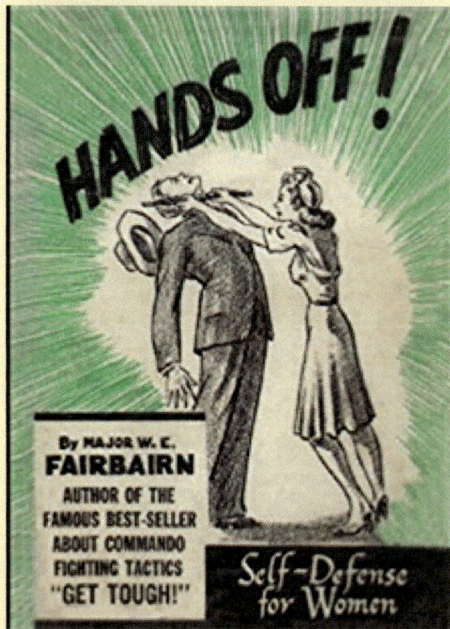

In 1907, William E. Fairbairn was 'drafted' into the Shanghai Municipal Police, at a time when the oriental port was an International Settlement, under the dominating presence of Britian, United States, China, and Japan. Then, Shanghai was the world commercial center, a wealthy city, and teeming port. It was also an extremely dangerous place, home to drug smuggling, slave trades, sex-trafficking, and ruthless, competing Chinese 'gangs' like the Tong. The Britons established a police force to maintain order.

Fairbairn and his Shanghai colleagues, such as the esteemed Eric Sykes, developed a self-defense system for the municipal police that was borrowed in large measure from Japanese Jiu Jitsu. Its effectiveness is described in the 1915 *Shanghai Municipal Police's Manual of Self Defense,* written by Fairbairn:

"In Jui Jitsu much pain is inflicted by means of light blows on tender parts of the body by pressure applied to nerve-centres and leverage applied to limb joints".

At the onset of World War II, Major Fairbairn was hired to teach these same techniques to American, British, and Canadian forces. By then, he was a formidable hands-on teacher and respected author. His 1942 book, *Hands Off!* was written for British women at home during WW2, when the English had their back to the wall, and were highly suspicious of enemy infiltrators. Being part of "the war effort" meant that everyone learned basic self defense.

The Unarmed Self Defense section in *At Home in the Real World*, is deeply rooted in the works of Major Fairbairn and his American protégé Lt. Col. Rex Applegate. Over time, Applegate updated Fairbairn's Shanghai techniques (armed and unarmed) and taught them to American soldiers leaving for WW 2. They're still taught today. Col. Applegate's book *Kill or Get Killed* is a classic in defensive and offensive skills.

Top image: From the 1915 *Shanghai Municipal Police's Manual of Self Defense*, Fairbairn (on the right) demonstrates the Chin Jab. He drives the heel of his palm into the man's chin by thrusting power up through his legs (heels are off the floor) and leverages his force by holding onto the 'bad guy's' lower back.

Bottom image: The cover of Fairbairn's 1942 *Hands Off!*, a self-defense book for women. The cover shows a woman using an umbrella as an improvised weapon in a modified chin jab.

INNER BADASS 101
The Basics

HAVE A PLAN

The moment your home is breached is not the time to freeze or wonder what to do. Develop a plan that is tailored to the layout of your dwelling; what you're physically capable of doing; and, what you're willing to do. Every self-defense situation will be different. You need simple, effective responses. Think several through, visualize and practice them, then refine them. That's what your brain and body will do in a high stress encounter.

HAVE A BACKUP PLAN

Even well considered plans go awry, so anticipate various scenarios and develop backup plans. Sure, your weapon of choice may be next to your bed at night, but what if you're in the living room or basement of your apartment building when a bad guy enters? Consider your options ahead of time. Know your exits. Hit the car alarm on your keys. Have improvised 'weapons' in every room. Lock an intruder in a room or garage. Keep your backup plans simple. They have to be effective. Make sure in advance that any moving parts of your plan are in working order.

THINK DIFFERENTLY

Look anew at the spaces you live in, imagine responding to a doped-up intruder in your hallway, kitchen, or garage. Turn fear into energy to develop viable options. Self preservation may call for damaging or surprising action to the bad guy. Or escape. Or complete concealment. Second chances are rare in self defense. There is no talking it over with an intruder as shown on TV. You'll have short seconds to respond effectively. Have a plan, and backup plans. Run them through your head. A determined mindset is critical. Open up new brain space for this stuff.

VISUALIZE

Imagine the flight part of "fight or flight". See yourself running five blocks to a known safe place. (*Yep, running!*) How exactly are you going to get down the

> The hardest, most critical part for regular people to get, is moving unequivocally into an offensive mindest.
>
> – my warrior mentor

fire escape, reach the ground, and get to a safe destination? Now, do it for real, refine it, do it again. See yourself executing a palm strike to the base of your assailant's chin with all your might. As he loses balance, grab his family jewels and twist hard. These grim scenarios may never come to pass, but once you've found what you're willing to do, visualize possible sequences of your plans and work out the kinks. Work through fears or revulsions that arise. You can do what you allow yourself to do.

LEARN

Dump all your reasons why self defense won't be necessary. Lose the "It's not gonna happen to me" talk. Ditch the "I can't live my life thinking like this" excuse. Seek knowledge and practical disciplines from good teachers (those who empathize, empower, and move students beyond self-limiting beliefs). Learn how to get your body strong, and do the work. Learn close quarter combat techniques. Learn the proper handling and shooting of firearms. Learn when not to use pepper spray. Be responsible for your own personal safety. You may be your first responder.

PREPARE

Get the minimal gear you need, and know how to use it. Keep it in good working order. Teach your kids how the home and family defense systems work. Craft a plan together, prepare together. Options may include: a safe room to retreat to; a ladder to descend from a second-story window; a fire extinguisher turned onto an assailant's face. Find strategies that work for you. Practice and refine them. If a tool doesn't work for you, find one that does. Then another.

PRACTICE

Explain your scenarios and responses with every member of your household. Run practice drills with everyone under your roof. Do it quarterly. Make practice rewarding. In an emergency, you and your kids will do what you practiced.

GET STRONG

Speaking from 58 years of significant middle age excess, I took back my health, then realized how much strength I'd lost along the way. I'm getting that back too. Feeling great is a great feeling. Bonus: the look and air of confidence will have a predator looking elsewhere—you'll be too much trouble!

In today's society, it is as proper for a woman to study self-defense and develop the warrior mindset as it is for her to hold a job, apply makeup, change a flat tire, or breast-feed her baby.

– Jerry VanCook

COMMIT

Don't dally in doubt. Commit to what you're willing to do and do it. Ramp up from zero to something…and keep going! Don't let anyone talk you out of it.

DEVELOP MUSCLE MEMORY

Muscle memory is the result of practice and more practice. It's your body doing what you've visualized and trained for. Whether it's running the path to the neighbors at night, or solidly racking your pump shotgun. Practice so you can do it in the dark. When you're pumped on adrenaline, only your simplest movements, your gross motor skills, will be available to you.

BUILD CONFIDENCE

In the writing of this book, my stepdaughter asked me if I'm paranoid. I said, "No Kate. I'm doing this to protect the people in my home, and the life I love. I'm actually feeling more confident." Know the feeling. You're the only one who can give it to yourself.

USE REASONABLE FORCE

Home Defender, to be legally defensible, use reasonable force. This means: Know when to stop. Inflict no additional harm to a badly wounded intruder who's been rendered harmless. Do not shoot a retreating assailant in the back.

3

There are 3 survival responses: Fight, Flight, and Freeze. The latter means an intentional act of being silent and not moving; or, concealment.

INNER BADASS 201
Adrenaline & You

Before we get into basic nuts & bolts of self defense in the home, you need to know about adrenaline: it may, or may not, change what you are capable of doing. As soon as your body perceives danger or registers an emergency situation, adrenaline hormones are pumped into your bloodstream and prime your body in a high stress second for Fight or Flight.

Professional soldiers who train constantly (at times, under extreme stress) develop muscle memory and work through the effects of adrenaline. They know what their response is supposed to be and they deliver. Planning and practicing helps us lay people too. We should be aware ahead of time what could happen in an adrenaline-fueled body. You'll see why it's very important to keep your response plans as simple as possible.

WHAT ADRENALINE CAN DO TO YOU

Here's how the Society for Endicrinology describes what can happen: "Key actions of adrenaline include increasing heart rate, increasing blood pressure, expanding air passages of the lungs, enlarging the pupils in the eyes, redistributing blood to the muscles and altering the body's metabolism, so as to maximise blood glucose levels [to supply emergency fuel to the brain-SM]".[3]

They also publish a list of potential physiological changes caused by adrenaline. (I added specifics). For reasons like age, health, and training, every person experiences adrenaline differently. However it affects you, know that it's short lived, and can last between 5-30 seconds. Here's what may, or may not, happen to you:

- Strength increases.

- Time slows down.

- You see only what's in front of you, from four feet onward. (You have no peripheral vision. You're oblivious to family members or additional threats standing off to the side. You may not see your gun's sights).

- Distance vision comes into greater focus.

- Eyes dilate, letting in more light.

- Hearing diminishes. ('Auditory exclusion', not hearing what's going on around you, not your partner, kids, barking dogs, gunfire.)

- Hands may shake uncontrollably, compromising fine motor skills. (Not being able to turn car keys in the ignition, dial 9-1-1, open a locked gun box, or pull a trigger).

- Pain doesn't register.

- Complex brain simplifies. (Solving new problems or making complex decision is out the window. Your body will do what you've trained it to do.)

WHEN IT'S OVER...
THE ADENALINE DUMP

When the adrenaline rush is over, weird things can happen. You may have gotten hurt in the encounter. Now, you notice you're bleeding. Your hands or knees may shake uncontrollably. A woman or man may tremble or sob; pee her/his pants; or vomit. Whatever the body does, it's Okay. When the threat is over, adrenaline has to physically express itself, and get "dumped" out of your body so it can get back to normal. There are no judgment calls. Trust your body's intelligence. If you're bleeding, get help.

Keep your plan and backup plans simple.

Think them through ahead of time.

Practice.

— Calamity May

INNER BADASS 301
Advice from Two Coaches

I've been fortunate enough to have two mentors for this section.

One is retired from a long, distinguished career in law enforcement; the other is a retired warrior, biomechanics expert, and trainer of elite forces. This is their advice to women, men, and anyone in a tough spot:

1. DE-ESCALATE AND DEFUSE WHEN POSSIBLE

There may be times when you're at an event or someone else's home where a person with attitude moves into your "zone". De-escalate the situation first: apologize if you bumped into him. If they move in uncomfortably close, tell them loudly and firmly "Back off", "Leave me alone", "Get out of my space". Attract attention, and he may go away. Or distract him by calling out to a non-existent friend. This is no time to be polite. If the situation worsens, get out or get ready.

2. WHEN YOU PERCEIVE A THREAT, COUNTER IT

As soon as a guy corners you, backs you up against a wall at a party, work, or behind an Army barrack, you'll know exactly what's up. Don't wait for him to raise a hand or smooth talk you into going somewhere else. The nano-second you detect physical violence, when you see the stance, the beginning of a grab, or fist clench, "Intercept the intent" as the famed Bruce Lee said. Strike a blow from your power center. Get him off balance—psychologically and physically. "Let your inner bitch fly!" as Dr. Ruthless says, a renowned educator in women's self defense. Don't wait for them to land a move. Strike first. Surprise is critical.

3. WHEN THE FIGHT COMES TO YOU, GO ON OFFENSE

When someone breaks into your home, or if you're attacked on the street, immediately go on the Offensive. Move in, strike, get him/her off balance. Disable the aggressor. Be relentless until they give up, give out, or run.

4. MOVE WITH SPEED TO SURPRISE AND DISABLE

Engage with total commitment. With speed, and intensity, yell or grunt, and get intense as you deliver disabling strikes to strategic places, no matter his size.

Watch the hands, not the eyes.

Eyes can't hurt you.

—Steve Barron

5. KNOW WHEN AND HOW TO BACK OFF

If the assailant is leaving your house, let him. Do not detain. Do not inflict harm from behind.

If the assailant is on the ground immobilized in your home, stay vigilant.

If armed with a gun or fire extinguisher, don't use it unless they come after you again. Call the police. Get help, if you can. Keep him in check until police arrive. (Tell the 9-1-1 dispatcher this is a crime in progress).

INNER BADASS 401
Fundamentals of Technique

If you're hitting the target and not falling down in the process, you're doing it right.

– Jerry VanCook

Mental Balance

For women and men who are smaller and weaker than 'the bad guy', balance and surprise are your allies. The mental balance of a bad guy can be upset by the surprise of an attack. Use any strategy he doesn't expect… a face smear, a fake out, call out to a friend who's not there….while he's re-orienting, take offensive action, take out his physical balance.

Work on it:
• It sounds easy to say, but in the heat of a bad moment, Keep Your Head. Don't melt or flip out with fear!
• Retain your mental and physical stability, rob your assailant of his.
• You may be the only who can save you.

Physical Balance

A person stable on their feet and able to retain balance can access her/his full power. If caught off center, one will first struggle to regain balance, then fight back. Don't let that be you. Balance is key.

In a personal battle, get your assailant off balance and keep them that way. Always maintain yours, learn to recover it quickly.

Work on it:
Build balance by standing on one leg. If needed, use one finger on a table or wall to stabilize yourself. Work up to a minute and more on each leg. After a few times, bend your knee and slowly lower yourself. Add time and depth to this exercise. When your leg starts to shake, hold for another 15 seconds. This cultivates physical balance, strengthens ankles and legs muscles, and improves overall blood vessel circulation.
• Dance, swirl, or spin often.
• Jump on a rebounder (a small trampoline) for 3 minutes daily.

See calamitymay.com for Fitness exercises.

Strength & Power

A worthwhile strike to a target inflicts pain, destabilizes, and disables. This is why you work on strength training. Core strength fuels one's abilities to build power, execute effective moves, and endure. Think of the self confidence!

Work on it:
• Practice High Intensity Training (aka Super Slow Strength Training) to develop core strength. It's basic and effective. No matter your age or physique, workout 1x/week for 15 minutes. Use your own body weight (no gym, no cost, no hassle).
• Strengthen your fingers, thumb, hands and wrists. Squeeze hand springs or a field hockey ball for 10 seconds then release over 10 seconds. I do slow wrist curls with a small steel bar to strengthen my wrists.

See how I do it, go to calamitymay. com and check out my Fitness blog posts.
• Punch a punching bag or strike a dummy (not your sibling!) Work on placement and accuracy. Speed will come. Always maintain your balance as you develop power.

Speed & Accuracy

Speed comes with training and preparedness. Know a few moves well. Have a weapon at the ready; or, be able to run like the wind. As soon as you sense imminent hostility, act without hesitation.

Accuracy. Make what you do count. The energy you expend in the first few seconds can save your life. Second chances are rare. Execute actions with the greatest accuracy you can bring to it. Whether it's a kick to the knee cap, unleashing pepper spray, or shooting a firearm—you have to hit your target. To hit your target in a moment of crisis, you have to practice before this decisive moment.

Work on it:
• At home, suspend tennis balls or improvise targets for hand and foot strikes. At a gym, use punching and kicking bags. Take Krav Maga or a self defense program that emphasizes close quarters combat.
• Find an empty lot, put up a paper target, and try out a 5-pound fire extinguisher. Try out pepper spray. Have these tools as part of your arsenal in the home. Practice with your tools. Develop accuracy then speed.
• Take classes in defensive shooting. It's different than target shooting.

Victims of assault are victims not because they lack the capacity to win fights but because they have been given absolutely no preparation to cope with this special kind of emergency.

– Bruce Tegner

SELF DEFENSE TOOLS:
10 Body Weapons

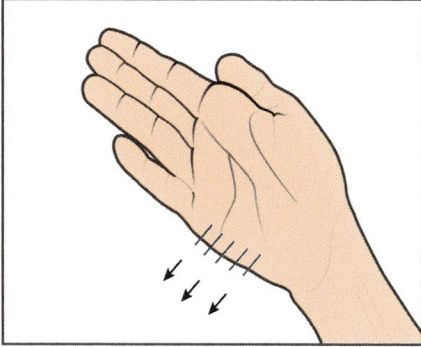

HAND
Edge of Hand, Blade

HAND
Heel of Palm

HAND
Double Fist

The Form: Strike with the outer pad of your hand, the fatty part. Keep fingers tight together and thumb firmly to the side. The power behind a vertical or diagonal blade, comes from the shoulder, down the biceps and through the forearm.

The Move: The most power comes from a horizontal strike palm down. It works well diagonally or vertically too. Force, speed, and accurate placement make blade strikes effective. Practice on something firm that gives, like a pillow or punching bag to get a feel for it. Knowing where nerve centers are located takes the 'blade' to the next level.

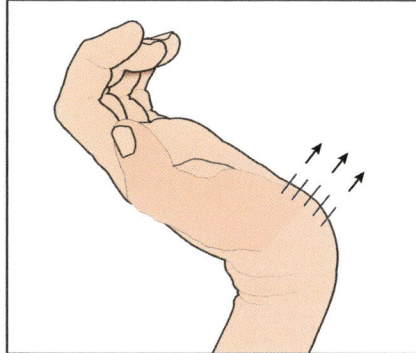

The Form: Your whole hand bends back at the wrist, almost 90 degrees. This exposes the 'heel' as the point of contact. It's positioned right over the bones of your forearm.

The Move: This is a forceful upward strike, one that can be to your advantage if you're short, and the attacker's tall. Force starts with an upward thrust of your knees. then shoulders and whole forearm. Strike to the underside of the chin, or the bottom of the nose, (the nostrils) (See page 102). Follow through. A nose strike in this way may do light damage (make eyes water) or do far worse damage. It's possible but rare when a civilian hits the nose with enough force to dislodge it into the brain. If it happens, it can be deadly.

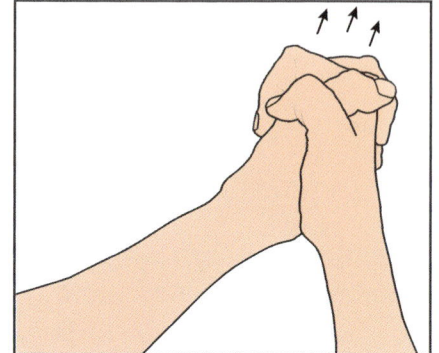

The Form: This move is a primary defense and offense move in Bulldog Jeet Kune Do, developed by the late Bob Chapman.

Both hands are clenched tightly into one fist. The fingers are not intertwined, instead, one hand 'folds' over the other, creating a tight hold.

The Move: This is a versatile move that can be used close in to strike the Adam's Apple. The arms deliver the force of the blow. The forearms are used to provide protection from incoming strikes. When clenched, the double fist is strong, it's hard to break apart. Get a tight hold and try to pull it apart!

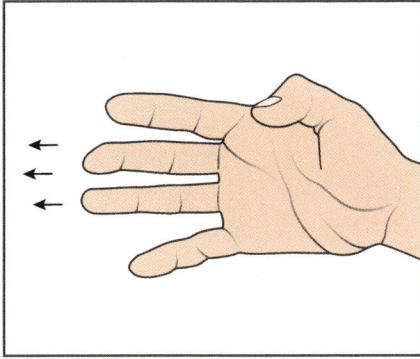

HAND
Four Finger Strike / Eye Strike

The Form: The four fingers of your strong hand are held rigid and slightly apart. The thumb is tucked to the side. It's best to practice with both hands in case one's disabled.

The Move: Palm down, this strike is a deep thrust into the eyes. Force is generated through your whole arm and travels through your stiffened fingers. Strike fast, with force, up close. The intent: go deep into the eye(s) and inflict significant pain and cause distraction. Out of four fingers, one's bound to hit into an eye. The assailant's hands will naturally go to his eyes leaving openings to strike.

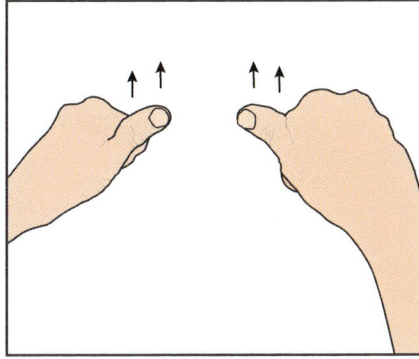

HAND
Eye Gouge

The Form: Both thumbs are out and rigid. The other four fingers on each hand can be used to hold onto his hair or ears to stablize the head.

The Move The intent is to penetrate deep into the assailant's eyes. If he's on top of you, leverage his head to make this successful. Grip his head/ears with both hands reserving your thumbs to gouge out his eyes. Plunge both thumbs into the eyes, go deep, push outward to the sides. You may dislodge one or both eyes. This is highly likely to affect a release to escape, or create an opening to strike. This is a close in move. Don't attempt at arm's length.

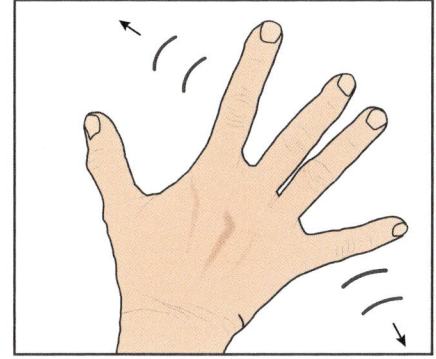

HAND
Open Palm / Face Smear

The Form: Open the whole palm of your hand, rotate your wrist in a back and forth motion. Make a quick, 'smearing' contact with his face, in the area of nose and eyes.

The Move: To smear a person's face is a distracting move. Everyone hates a sweaty palm rubbed into their face. Get close, keep it light and brief. They're likely to pull back. You just bought space. Don't overextend your arm, it could be grabbed. This move should be unexpected and surprising. Prepare to followup with different moves based on openings. (Used by Grace Kelly in the movie, High Noon! See insert page 117).

A dame that knows the ropes isn't likely to get tied up.

—Mae West

10 Body Weapons, continued

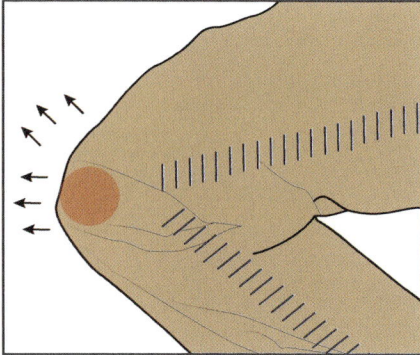

KNEE
Knee Jab

The Form: Armored with a solid knee cap, a knee jab is delivered up close, with force. If you can, hold onto the bad guy's shoulder while delivering the jab.

The Move: The Vagus Nerve runs the outside length of the leg. The outside seam of a pant leg is a pretty good guide. Use a knee jab to strike there. Hold onto bad guy if possible. A jab can also work up and into the soft stomach area.

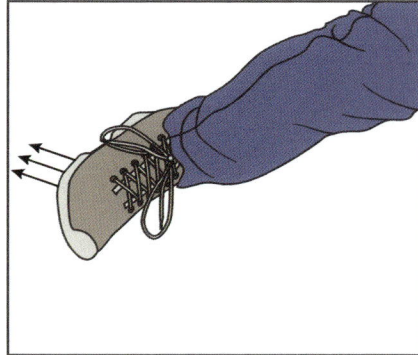

FEET
Outer Edge of Foot

The Form: A forceful kick with the outside edge of your foot, or the sole of your foot can bend a knee in a direction it wasn't designed to go. Never kick with your toes.

Move 1: Strike the bad guy's knee straight on, or sideways. You, however, need to regain balance quickly.

Move 2. Shin Kick: Kick with outer foot to top of shin (below kneecap), and bearing weight, drag your foot down along the shin (which has lots of nerves). Finish with a Foot Stomp.

FEET
Foot Stomp

The Form: The heel of your foot, is forcefully stomped down onto the assailant's foot, preferably the instep.

The Move: 26 bones are in each foot, many can be broken with a forceful, downward stomp to the instep, or top of foot.
Use a foot stomp at the bottom of a shin kick. (See Outer Edge of Foot Move 2).
High heels? If you didn't kick them off by now, Go for it! Plant that stiletto! Flats? Sandals? Work boots? Drive down hard with leg power.

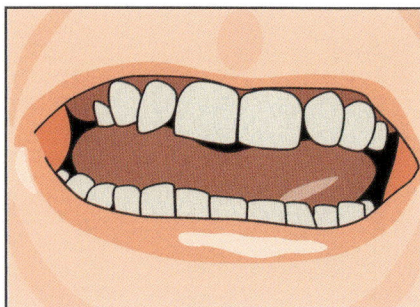

TEETH
Biting

The Form: Bite with teeth.
The Move: This is an ancient tactic from the book of personal war. Use the incredible biting down power of your incisors to rip flesh and take off ear lobes, chunks of neck.... If you're in that close, you have to inflict physical and psychological pain.

SELF DEFENSE TOOLS:
15 Targets

Now that you have a sense of the 'weapons' you carry daily, here are places to strike. These are basic, time-tested moves available to laypeople of average strength and ability. **Though I've written as clearly as possible, nothing takes the place of professional instruction.** Take a class in close quarter combat, Krav Maga or Bulldog Jeet Kune Do. Before your first class, ask to be shown proper striking forms, so you don't hurt yourself. (Or you may not go back!) With practice and repetition, you can develop accuracy, speed, and force. You want muscle memory. This is unarmed self defense in the most dire of circumstances. Some moves can be lethal, though rare.

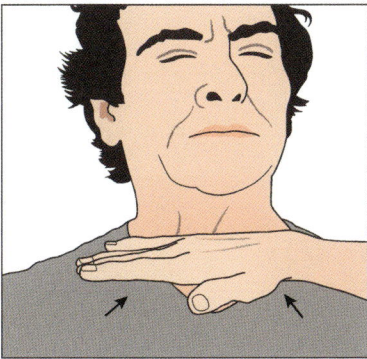

Blade to Wind Pipe

The Form: Using a rigid Edge of Hand, or Blade, deliver a side strike to the Adams Apple (shown) or Jugular vein on side of neck.

The Move: Palm down, outer edge of hand (the fatty part) strikes with a forceful, lateral motion. Look for the next opening.

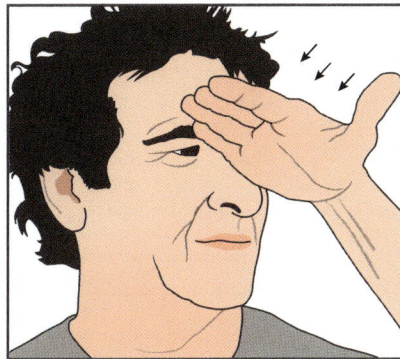

Blade to Nose Bridge

The Form: Same rigid Edge of Hand, or Blade, delivered to the nose bridge.

The Move: This will only work if you're taller than your target, or if they're within lateral reach.

In the image above, my brother's taller than me, and I'm reaching too high, exposing my torso. (Not advisable).

Potentially lethal.

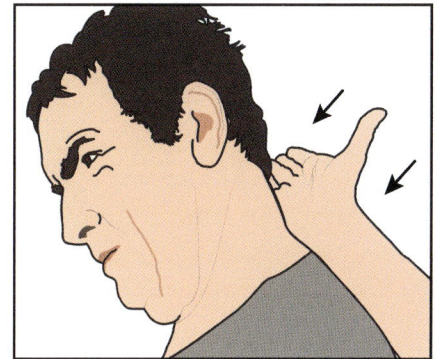

Blade to Nape of Neck

The Form: Same rigid Edge of Hand, or Blade, delivered to the base of neck, above the last vertabrae.

The Move: Again, this will only work if you're taller than your target, or if he's bent over forward. If they have a thick neck, it's likely to have little effect.

Hands are faster than feet and strike smaller targets.
Legs deliver more power through kicks and knee jabs.
— warrior mentor

15 Targets, continued

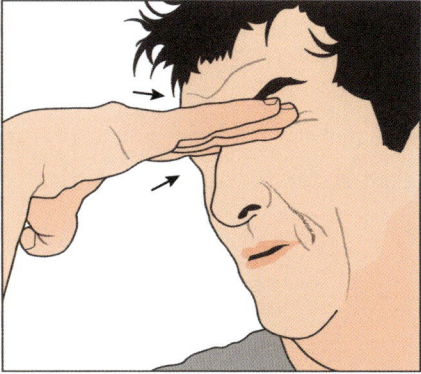

Finger Strike to Eyes

The Form: Your hand is flat, palm down. Fingers and wrist are rigid; fingers spread open.

The Move: Thrust all four fingers with great force into one or both eyes. Strike with all four, but only 1 or 2 are likely to get into the eyes. "The intent is to drive all four fingers into both eye sockets. The index and middle fingers are destined to go in".-Mendell

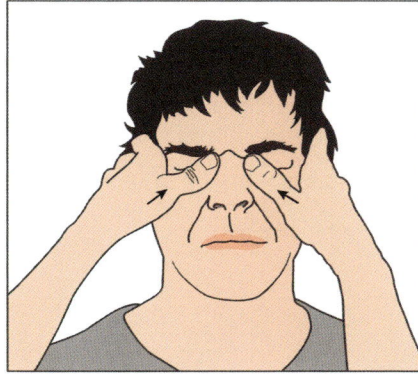

Eye Gouge

The Form: With rigid thumbs, plunge both thumbs to the inside of the eye socket and exert force out to the side. Your fingers can be used to stabilize an attacker's head by gripping their ears, or rear portion of their head.

The Move: This is a close-up move. A guy on top of you move. The intent is to gouge out his eyes. Yeah, gross. But so is what he would've done to you…. Do not attempt this at extended arm's length. To do so expresses intent, leaves your chest open, and exposes your arms to grab and twist, or break.

Possible Next Moves: Ear Slap; Blade to Jugular, Groin Kick; Knee Kick

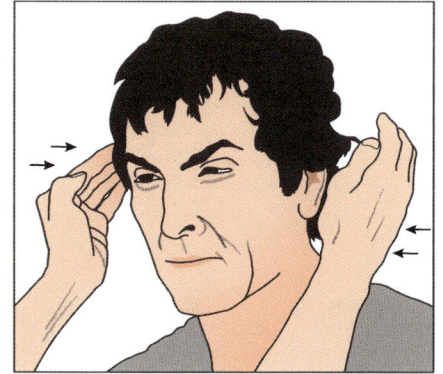

Ear Slap

The Form: Both hands straight out or lightly cupped.

The Move: Slap both ears at the same time from front or behind. This can cause intense pain; disorientation; or, eardrum damage. A concussion is possible. The assailant will let go, and grab their ears or head. There's bound to be openings….do some damage while their hands are up…

Possible Next Moves: Eye Strike, Chin Jab with extension; Groin Kick; Groin Grab & Twist; Foot Stomp

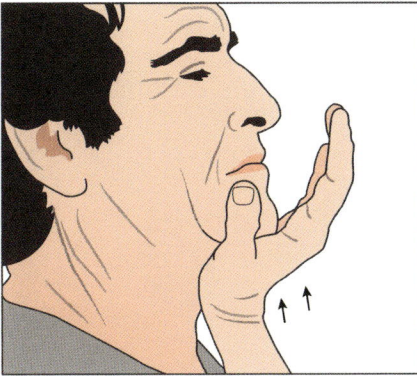

Chin Jab

The Form: The wrist is all the way back; fingers are back and curled. This is a good move for a short people against a taller assailant.

Move 1: Drive the heel of your palm forcefully up into an attacker's chin with all your might, everything you've got. "The heel of the palm is backed by the full weight and strength of the body where power and strength are demanded."-Mendell

Extension: Following the chin jab, extend your fingers up and strike deep into the eyes.

Potentially lethal.

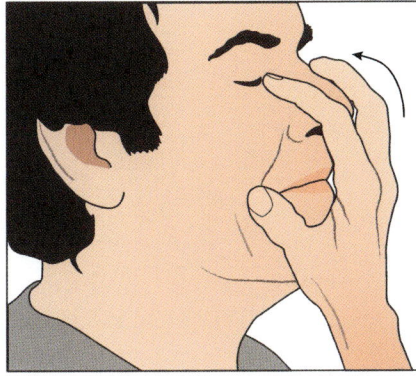

Chin Jab, extension

Extension: After the chin jab, If you have the reach (long fingers help), follow through with an eye strike. Extend your fingers up and strike deep into the eyes. Jab and eye strike takes place in one fast, seamless movement.

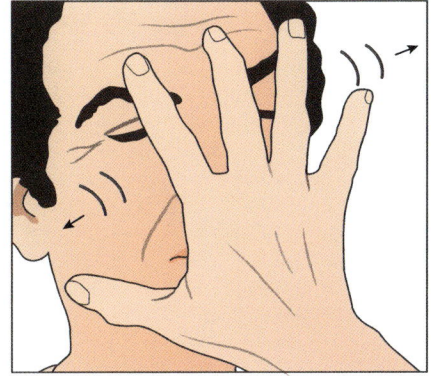

Face Smear

The Form: The inside of your hand, the palm, is wide open.

The Move: Who likes to get their face smeared with somebody's sweaty palm? Nobody! Smear your sweaty palm in a bad guy's face, and he/she's likely to pull back. It's a move that distracts and surprises. For a half second, the bad guy will be mentally off balance. This buys you a half second to find an opening for your next move, or escape.

Always look for the openings.

— warrior mentor

15 Targets, continued

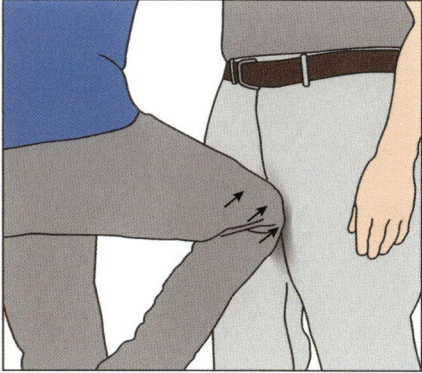

Groin Kick

The Form: Deliver a sharp knee jab upward or straight into the target's groin (crotch).

The Move: Drive a strong knee jab up or forward into the testicles. Really deliver it. (If you play darts, think 'follow through'). Most, but not all men, are super-sensitive here, be forewarned!

Next Move: When the bad guy leans forward to grab his balls, intercept his face with a Chin Jab, or, Blade to Back of the Neck. Or kick his knee cap backwards to demobilize, and escape.

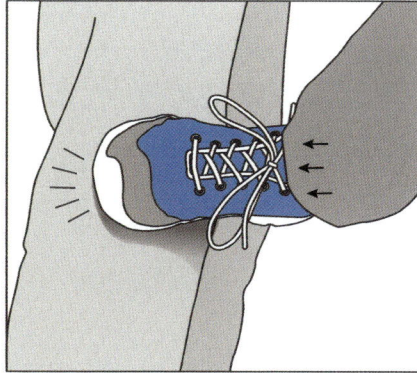

Kick to Knee Cap

The Form: Stand sideways on one leg, and kick with the other. Use outer edge of your shoe, or sole of your foot, to kick his knee cap backwards, or sideways.

The Move: Be at an angle where you can center on one leg and deliver a strike with the other. To kick in a kneecap, get your distance, angle and power right. This is a very effective move to disable an attacker, especially one taller than you. If well landed, expect dislocation, hyperextension, or a broken patella.

If you can't kick with force and remain standing, don't do it.

Shin Kick & Drag

The Form: With the outside edge of your foot, kick below the knee-cap at the top of shin.

The Move: That bone on the front of your calf is called a shin—it's loaded with veins and nerves. With the outer edge of your shoe, sharply kick the bad guy at the top of shin (just below his knee cap) and apply pressure while dragging your foot down the length of the shin. Painful! Follow with Stomp to Instep (next). Most effective in shoes that have a stiff edge.

In the 1952 movie High Noon, Amy Kane (played by Grace Kelly), was held in a tight waist hold from behind by bad guy Frank Miller. In rapid sequence, she wriggled out of the hold, used a Face Smear, and followed with an Eye Gouge. In doing so, she freed herselft and distracted Miller from shooting her husband Marshal Will Kane (Gary Cooper).[4]

Foot Stomp to Instep

The Form: Heel of your foot, stomped down onto the top of assailant's foot or instep (where foot meets ankle)

The Move: Forcefully stomp on top of the foot. (Small bones may break).

It's a good finishing move at the bottom of a Shin Kick (left). Wearing high heels, plant that stiletto! Flats? Sandals? Work boots? Drive down hard with your heel using all your weight. Don't fall down in the process. Stabilize yourself if needed.

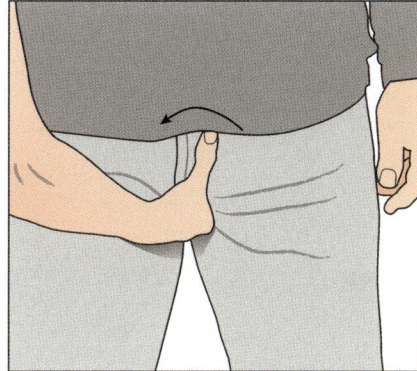

Grab, Squeeze & Twist Testicles

The Form: You read that right! If in close, reach up into the groin with a free hand, grab his testicles hard, squeeze, and twist.

The Move: Seriously, if the situation dictates, use your strong hand (or free hand) or to grab, squeeze, and twist his family jewels. Twisting is key and makes this move highly effective and temporarily disabling. Getting through denim may be challenging. Make it count! Execute with 100% of Primal You.

You just bought some time to flee or execute a few moves to slow down the bad guy.

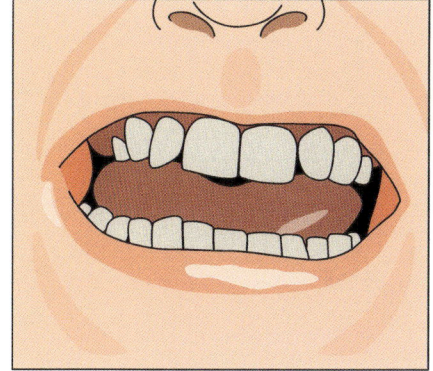

Bite Flesh or Body Part

The Form: You're right up against your target, you bare your teeth, biting down with your mighty jaw power.

The Move: "Whaaat?!" I can hear it now. Like the animals that we are, a powerful, deep bite can rip flesh and draw blood. It's what inmates do and take off ear lobes and chunks of flesh. It's highly effective and psychologically devastating to the opposition. You have to be in very close, and desperate to make it count. This is serious business, not a sandwich: Bite down with great force as if your life depends on it—it just might.

ZONE 3 - SELF DEFENSE CATALOG

Practical self-defense techniques are those which depend to the least degree upon relative power, size and build. Kicking into the knee or shin is an example of a highly practical self-defense action.

– Bruce Tegner

If it's stupid
but works,
it isn't stupid,

— Murphy's Law
of Combat

ARMED DEFENSE:
Improvised Weapons

Almost anything can be an improvised weapon. In the 1976 book, *Black Tiger Kung Fu*, household "weapons" common in China for millennia included rakes, hoes, benches, umbrellas, walking sticks, even baskets.

Your home surely contains household objects that can be used as tools for self defense:

- hot coffee thrown in the eyes

- car key's panic button to distract and draw attention

- car keys as physical weapon to scratch, cut, and draw blood

- cell phone as striking weapon (corner to temple or top of head)

- handle of an umbrella or cane to strike deep into soft belly or thrust into an Adam's Apple

- cast iron frying pan into a face

- knives, old-fashioned knife sharpener, scissors...you get the picture!

First, if you can, distract. Set off a home or car alarm; or, throw something into their face. Aim for the eyes. Be ornery, be a lot of trouble. Buy time to escape if you can. If you can't, reach for your preferred self-defense tool that you put in a strategic position. (If there is absolutely nothing within reach, see pages 112-119).

Remember: good placement + force = pain and/or reduced movement. If they don't leave, run when you can, or take a stand. Follow through 100%. Don't do anything half-heartedly.

Look around the room you're sitting in right now. What improvised weapons do you see you right now? Get creative. What are you able to use? What are you willing to use? (Re-visit the Willingness Exercise). While you're at it, look at your exits, can you get out quickly?

One more thing: in addition to physical tools, use psychological force. Psyche out a bad guy with yelling or war cries. Go ballistic! Unleash Primal You!

SHARKIE & BOLT

Brian Keith offers, "Improvised weapons are essential. I carry a very large 'Sharkie' that I bought online for $6. When I flew to New Orleans last November, I carried a 5/8" wide, 6" long stainless bolt.

If you hit someone full force in the ribcage with a 'Sharkie', or bolt, they're going to be hurt. It concentrates force to a very small area.

Sometimes I carry a stick to use as a Kuboton, (a hand-held striking instrument). Knife and stick training is great because most knife and stick techniques work with everyday items."

I travelled with a 4" bolt in my carry-on luggage, When it was held up for inspection by Mexican security, I said, "I'm going to Acapulco. I hear it's dangerous". They let me leave with it. I carried it with me constantly.

FIRE EXTINGUISHER

Do you have a fire extinguisher or three in your house? I hope so.

Not only can you use a fire extinguisher to put out stovetop fires, or wastebaskets flambé, but in a pinch, they make a handy self-defense weapon. Home fire extinguishers typically have an ABC rating, good for putting out fires involving paper & wood, flammable liquids (grease), and electric sources.

Make sure the dry chemical in yours is monoammonium phosphate (MAP). If you use sodium bicarbonate in a defensive moment, you'd only coat the bad guy in baking soda.

The effects of MAP aren't debilitating, they are however, irritating and distracting! The Material Safety Data Sheet spells out the short-lived, non-lethal, hurt MAP can put on a bad guy:

- May cause eye irritation.
- Slightly irritating to skin.
- Dusts may cause upper respiratory tract irritation (very bad for asthmatics)
- If ingested, may cause abdominal cramps, nausea, vomiting, diarrhea.

When you spray his face, his hands'll likely go up to protect his face. You just bought a half second. Look for the openings. Consider a groin strike, or reach for the next kitchen tool at hand. Or run, if you can.

Pitcairn Gun Club, features an aging display of 'assault weapons' found in most homes:
• umbrella
• frying pan
• hammer
• rolling pin
• broom
• golf club
• softball
• brick
• walking stick
These may have been effectively wielded by your foremothers a long, long time ago!

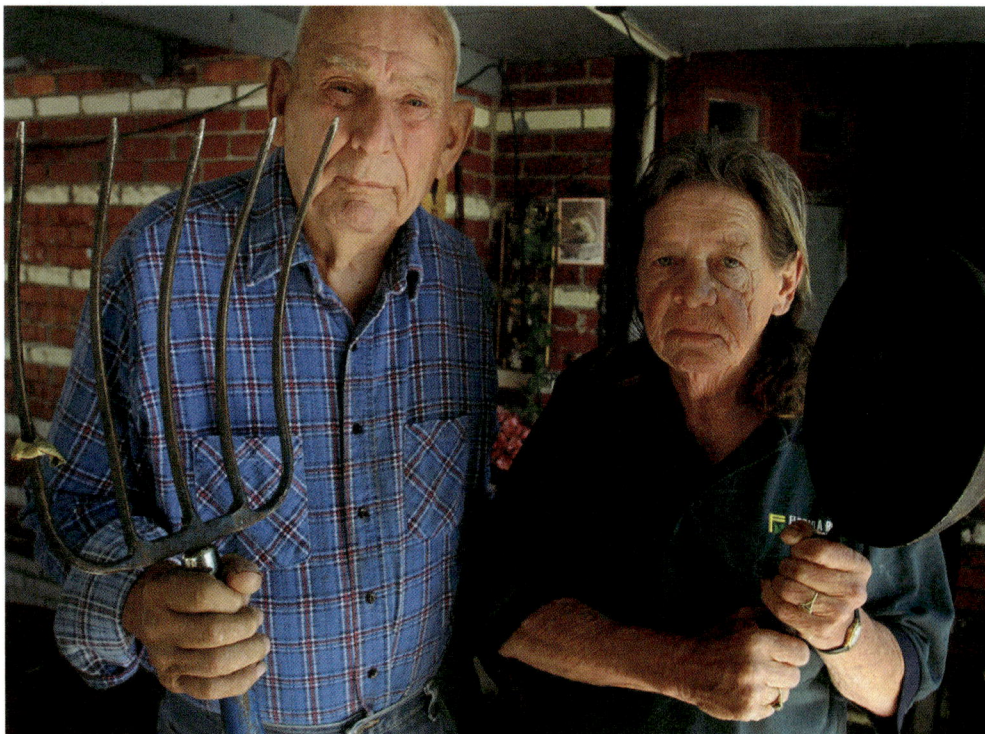

"The elderly gentleman that you see in this picture is 81-year-old Bobby Smith; the woman next to him is his 65-year-old caretaker Luvina. The pair recently made mincemeat of a man armed with a gun who entered their home and demanded money".[5] Bobby, a Korean War vet hit him in the face with hot frying pan, then stuck him in the side with a pitchfork. The bad guy ran out, wounded". A footnote worth reading!

Let your inner bitch fly!

— Melissa Saolt, aka Dr. Ruthless

MINDSET

Melissa Soalt, aka Dr. Ruthless is my favorite self defense expert. This is an excerpt from an article she wrote on self defense, mindset, and improvised weapons. The article, *"Improvised Self-Defense Weapons: How to Turn Everyday Objects to Your Advantage"* appeared in Black Belt's online magazine, blackbeltmag.com, date: July 9, 2012.

"Betty Jo is home alone, wearing her favorite flannel nightgown. She shuffles into her U-shaped kitchen and fixes herself a cup of Sweet Dreams tea. Suddenly, the kitchen door is kicked in and the prospect of sweet dreams turns into her worst nightmare. "Shut up! Shut the f*** up!" the hulking man spews as he closes in. Fearing for her life, Betty Jo backpedals in horror, becoming trapped in a corner. The attacker slaps and punches her, knocking her to the floor. The rest of Betty Jo's nightmare appears in the morning papers.

It didn't have to be this way, however. Quick thinking, savage instincts, a surly survival mindset and some basic self-defense moves coupled with some improvised self-defense weapons could have turned her nightmare into his horror story.

Let's replay this with a different ending, taking it from the moment the attacker enters: In spite of her terror, Betty Jo glances around, hunting for and maneuvering toward self-defense weapons of opportunity. She feigns weakness, pleading to buy time, but has already made up her mind: The only way out is through. Taking matters into her own hands, Betty Jo erupts like a fireball.

She grabs a nearby metal colander and whips it at his eyes. He flinches, and by the time he recovers, she has already snatched the boiling pot of water from the stove and thrown it in his face. Backed up by all her might and a bellicose war cry, she slams her handy cast-iron pot cover into his mug.

As his hands reach for his pained face, she assails him with knee strikes. In spite of fearing his counterattack, the wonder drug of adrenaline propels Betty Jo into some fierce self-defense moves. She grabs her attacker by the hair, smashing him face-first onto her granite countertop. She kicks his legs out from under him, grabs a knife from the counter and bolts out the door.

The morning paper reads: Betty Jo Goes Ballistic! Serial Rapist in Prison Hospital!

My version is dramatic and idealized. I don't mean to suggest that striking back is always the best or safest option, but it illustrates a crucial lesson: Self-defense moves are most effective when you can be adaptive. To be prepared, you must own your world and learn to transform everyday objects into self-defense weapons".

NON-LETHAL DEFENSE:
Pepper Spray

You don't need a trendy gizmo (like a cell phone app) for personal protection. What you need is a Primal You mindset, knowledge of a few basic moves, and a small, self-defense tool that you know how to use.

Here's one I really like. It's non-lethal, non-toxic, legal to use, and burns like hell: Fox Labs pepper spray.

WHY FOX LABS?

Their products feature a high concentration of extremely hot capsaicin oil (OC)—the active ingredient in hot peppers—5.3 million Scoville Heat Units, in fact. That, is smoldering hot. It's the most potent pepper spray sold. Its other features:

- You control the duration of each spray.

- OC concentrations vary 2% and 6% OC. (I buy the 6%).

- Canister sizes vary. I like the Jogger, (right).

- Some models have UV dye to mark the bad guy.

- Each canister's warrantied for 3 years from date of manufacture to be free of defects.

- "Inert" canisters, without hot stuff, are sold so you can practice. Buy several to get a feel for distance, spray pattern and dispersion.

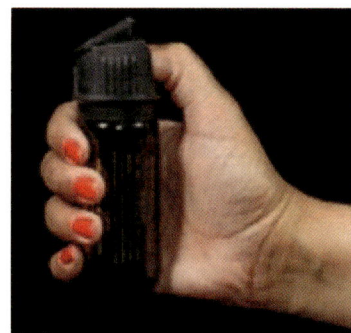

I've gotten hot pepper juice in my eyes about 15 times. It goes like this: I cut some jalapenos, add them to a pot of chile, forget about it, then rub my eye five minutes later. (*Yeah? You've done that too?!*) Well, Fox Labs' stuff is about 500x hotter than that! If you get good coverage on a bad guy's skin, face, eyes, it can disable him for 10-15 minutes. If he inhales it, he's out of commission even longer. It's legal to use and non-toxic.

CAUTION

A tiny percent of people exist in this world have trained to work through being sprayed with pepper spray; or, have experienced it often enough (rough prison inmates) that they keep going after being sprayed. Have a backup plan!

PEPPER SPRAY

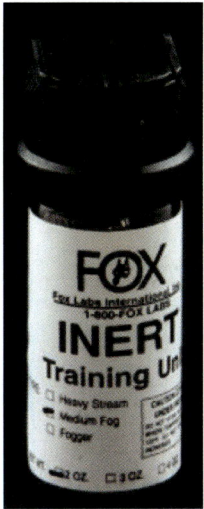

Practice! You can purchase inert pepper spray that mimics the stream or cone fog patterns. Try it outside. Experience what it does in light wind or any air movement. **Stand downwind!** See how far it throws. See the difference between cone and stream patterns. Feel what it's like to deliver a long burst, or several short ones. Note: even with the inert container, you may get a tiny percentage of active ingredient on you. Some people experience a tickling throat if they inhale it., others feel no effect.

On the Internet, are YouTube videos of people having it sprayed on them. There are also advocates suggesting having it sprayed on yourself to experience it. I don't intend to! MSRP, 1.5 oz. inert training unit,$15.99, Mean Green, 1.5 oz canister, 6%, $19.99

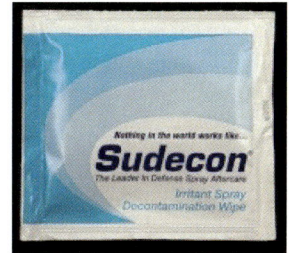

How Hot is Hot?

	Scoville Heat Units
Pure Capsaicin:	15,000,000 - 16,000,000
Pepper Spray (Fox Labs):	2,000,000 - 5,300,000
Trinidad Scorpion:	1,200,000 - 2,000,000
Scotch Bonnet and Habanero:	100,000 - 350,000
Thai Pepper:	50,000 - 100,000
Cayenne:	30,000 - 50,000
Jalapeno:	2,500 - 8,000

(Source: http://www.pepperscale.com/)

Wipe it off! Use Sudecon's decontamination wipes, if you come in contact with any kind of pepper spray. The manufacturer states, "It strips chemical agents from the skin instantly, takes away the burn, allowing the eyes to open in just 7 to 15 minutes or less". Keep 1 or 2 in the same place as your canister. **MSRP, $1.75 each**

Learn! FoxLabs.com is a very informative web site.

Caution! Do not use if you have asthma or any respiratory illness. If used indoors, you may come in contact with surfaces or people sprayed with it. You may even inhale it. Though it's non-lethal, it can adversely impact you for 30-45 minutes. You must learn where, how, and when to use it. Taking a class in the use of pepper spray is recommended.

Get 15% off your order, (same discount police departments get). Go to calamitymay.com, search for "Fox Labs". At Fox Labs checkout, enter promo code 'calamity'.

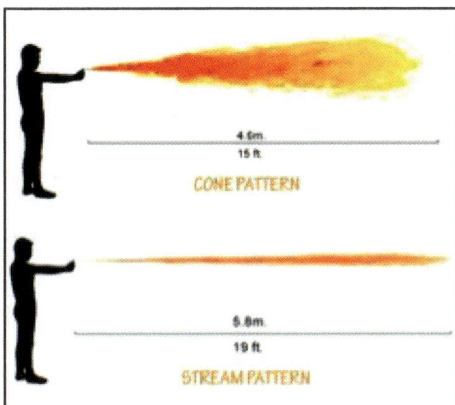

Fox Labs recommends cone spray for civilians.

Read how a young lady I know in New Orleans successfully defended herself using Fox Labs pepper spray, at calamitymay.com. Search "pepper spray".

HARD QUESTIONS TO ASK ONESELF

Defense of self, children, the elderly and disabled in our care seems natural to many. But if you're new to considering firearms for self defense, you need to ask and answer, sobering questions. Even if you've logged 1,000 hours of range time, revisiting them is no waste of time. Consider the following questions, then use your heart, gut, and brain, to arrive at your answers:

Am I willing to point my gun at a human being who I perceive to be a serious, imminent threat and pull the trigger?

Can I comprehend the idea of shooting one or more rounds into that person until he is stopped or in retreat?

Would I be able to live with potential emotional scars, and social and legal ramifications of having shot or killed a person?

A Well Armed Woman blog post wraps up this line of thinking this way: "If you are not willing to explore these questions and cannot answer yes...you are not ready and should not own a gun. To do so, would put you and others at great risk. Consider perhaps other less-lethal forms of self defense".[6] I agree.

If you're willing to explore the idea, read on. If it's not for you, consider all the options in Unarmed Self Defense, Improvised Weapons, and Non-lethal Pepper Spray. Consider your options for a safe room, concealment, or escape. Whatever you choose, practice.

If there is a breach of your castle walls, I want you to think very hard about how exactly you're going to stop someone coming at you from 20, 15, and 5 feet away. (That 20' distance can be covered in less than2 seconds if he's running). Even if you have a gun, if you don't know how to use it; can't access it inside of one second; aim and shoot it proficiently in the next, a gun alone may not make a difference.

GUNS & YOU

IF YOU DON'T TRUST YOURSELF TO USE A FIREARM	*GET TRAINING, OR STEER CLEAR.*
IF YOU'RE MORALLY OPPOSED	*FIND ANOTHER WAY.*
IF YOU'RE CLOSED TO EVEN CONSIDERING A GUN	*SKIP THIS SECTION.*
IF YOU'RE PRONE TO DEPRESSION, ANXIETY, PARANOIA	*FIND ANOTHER MEANS OF SELF DEFENSE.*
IF YOU "JOKINGLY" KID AROUND ABOUT SHOOTING YOUR HUSBAND IF YOU HAD A GUN	*GET A MARRIAGE COUNSELOR.*

Choosing to train, handle and use a firearm in your home is a commitment to being responsible, safe, and competent. Guns are tools—like cars and chainsaws—they're only potentially lethal when people operate them.

GUNS & ME
My Firearms Story

Shooting with friends and familiy.

At 17, with family members, I began plinking (basic target shooting) at paper targets, 8-track cartridges and/or defunct turntables that my brothers dragged in front of the tall dirt mound at our country home. Target shooting with shotguns and pellet guns was fun and allowed for friendly competition and the chance to better one's last shots. It got better....

On my 26th birthday, my boyfriend gave me a BB air rifle. (Some guys know the perfect gift!) I had so much fun target shooting from the patio of my secluded home that I took it inside. During a bathroom renovation, I had a 10" high platform built to elevate the claw foot bathtub to the same height as the window sill facing me (the bather). From this unique vantage point, I could soak in the tub and shoot my BB air rifle, window open of course, taking aim at raw eggs tied to trees branches 30' away (a steep hillside rose behind). If shooting from my patio was fun, it was magnitudes better from the bathtub!

Harry, a friend I knew from Explorer's Club, (an outdoor adventure group), asked me out on a date to shoot clay pigeons. He brought his father's Winchester double barrel, side by side. With direction, I lifted the hefty thing to my shoulders. He tossed the clays, I hit most of them. More than he did. A year later in Maine,

Lenn asked me out on a clay pigeon date. He brought his Browning A5 semi-automatic. Super smooth! I got most of my clays including two hanging in the air. I was thrilled, he wasn't. Neither asked me out again.

BB air rifles and shotguns, clay pigeons and plinking were fun, still are. Handguns though, had a different aura. I remember well the roar and fire that ripped out of my brother Kurt's .357 magnum at an indoor range 35 years ago. It rattled my bones and hurt my ears. Empty of rounds, he turned to me, "Here, Suz, take this and hold it just so...." Ugh. Suddenly, this wasn't fun anymore. Things were feeling potentially lethal. I was afraid of holding that gun and didn't like shooting it.

Fast forward to the present. Gunfire and high burglary rates in my neighborhood, led me to consider this question: *Am I safe in my home? Should I get a gun, just in case?* This thought set me to pacing. I called a friend from the past. Terry was as much a renaissance man as I'll ever know: novelist, African game hunter, blues guitarist, inventor, gun nut. At his kitchen table he made time to explain the critical role of firearms for self defense in the home. He suggested that I learn to shoot properly. "Okay", I said to myself as much as him, "I'll do it".

Today, my approach to firearms is surprisingly enthusiastic. What started with plinking evolved into NRA safety-infused training, and now, I'm an NRA-certified instructor. I practice shooting shotguns and handguns almost weekly.

My approach to firearms and life is informed by two other disciplines, tai chi and meditation. Because of them, I'm more grounded, relaxed, and calm. I also practice strength training weekly, focusing on core muscle groups and wrist, hand, and finger strength to better manage firearms and open reluctant pickle jars.

I enjoy teaching women to shoot. They take well to instruction, put safety before ego, and tend to be good shooters to boot! I understand the motivations behind self defense and share concerns about likely disparities in size, weight, and power of a bad guy versus a woman.

As a starting point however, I will always recommend preventing home break-ins in addition to the need for self defense. Let's be ready for both!

PROPER TRAINING FOR A BEGINNER

At a sportsmen's club near Pittsburgh, I signed up for (8) 2-hour lessons in Basic Pistol Shooting. The curriculum, designed by the National Rifle Association (NRA), is available nationwide. Lo and behold! The instruction was excellent; the instructor to student ratio high; the emphasis on safety, paramount; political zealotry, zero. The instructors were respectful, competent, and I felt safe. I loved every class.

My classmates and I (6 women and 4 men), received classroom instruction first, then entered the indoor range to learn to shoot long-barreled .22 target pistols. By the fourth class, we could bring our own handguns. By then I had a new-to-me Smith & Wesson 9mm semiautomatic and was getting used to its feel and fire. It's not a handsome thing, but I wield it effectively. I passed the NRA Basic test and was a regular at the indoor range that winter, firing about 1,500 rounds into lots of little black, concentric circles.

Target practice and self-defense shooting practice are two different things.

In the second NRA training course, Personal Protection Inside the Home, I was equally engaged in the classroom instruction and live fire practice. On the way home one night however, immobilized in tunnel-repair traffic, the purpose of this class dawned on me: to teach me how to shoot an intruder inside my home. No more black circles, raw eggs, or defunct stereos. This is about potentially killing someone, whose center of mass is not outlined on paper, but is flesh and blood with a beating heart. The satisfaction of that night's bullseyes drained away. I was sullen. I had to think hard and feel my way through this significant issue. I still do.

NRA's shooting for self defense classes give homeowners very specific instruction to encourage an intruder to leave. That's a great first option. In my mind, and allowable by Pennsylvania law, if the intruder doesn't leave, poses an imminent threat to me, or decides to come upstairs, he'll be warned once. I've made up my mind about what I'm willing to do. And I know what I can do under Pennsylvania law.

WHAT YOU PRACTICE IS WHAT YOU'LL DO The five mentors I've had the great privilege to learn from, all share the same advice: What you're willing to do, and practice, is what you'll do under stress.

The crux of self defense shooting is using a gun you know, handling it without thinking about it, placing shots accurately and fast, and using appropriate ammunition inside the home. Do your best to be prepared.

To help you learn this stuff, there are expensive schools and affordable ones. Seek the best training you can afford. Go once a year. Get a feel for defensive/offensive mindset and efficient techniques. If you're not going anywhere, shoot

what you have. Practice often. Get comfortable with your gun. Also, learn to clean your gun, you'll become familiar with it and have greater confidence shooting it.

SELF DEFENSE SHOOTING

Target shooting is fun, but it's not self defense shooting. There's a big difference.

For target practice, the following protocols are taught. The Weaver and Isosceles stances are stable ways to hold your body for target shooting. Both have you standing up straight and still, upper torso fully exposed. You take time to align the gun's back and front sights with the target; grip the gun properly, breathe calmly, and slowly press the trigger–all fine for target shooting. For self defense, those protocols are impractical, unrealistic, and don't take into account effects of adrenaline.

At home in the real world, self defense happens within seconds, at close range, and probably with moving targets.

In the grip of a potentially life-threatening intrusion in your home, a very different shooting style is called for. It's called point shooting. The reason it works for women and men, novices and soldiers (from World War II through today's Special Forces), is that point shooting stands up under extreme stress. Its whole premise is based on gross motor skills—what you're left with when your brain takes a hike and adrenaline takes over. Point shooting works when it's all you can do to retrieve your gun, hold it in a tight grip, move as needed, take aim without gun sights, and shoot quickly with both eyes open.

2 DAYS OF TRAINING

In October 2014, I took 16 hours of instruction in 'Applegate Point Shooting' taught by Stephen Barron, a foremost instructor in the field, at Hocking College in Ohio. The classroom instruction on the history of point

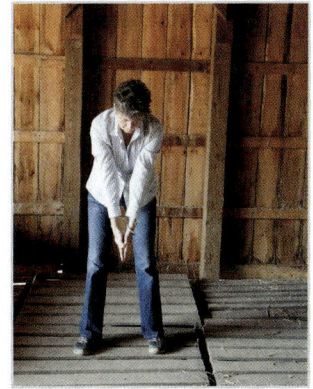

'Isoceles' means a triangle with 2 sides of equal length. The Isoceles stance calls for both arms extended out, and of even length. Knees are bent slightly. Classic target shooting stance. (Feet are always shoulder width apart).

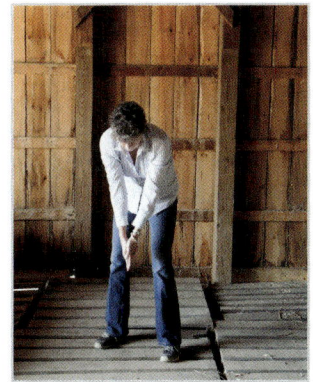

In the Weaver position, you push your strong arm (right arm on most people) forward, and at the same time, pull your support arm back, it will bend at the elbow. This tension creates a stable shooting platform. Feet can be even for target shooting. Knees are slightly bent for target practice; deeply bent for point shooting.

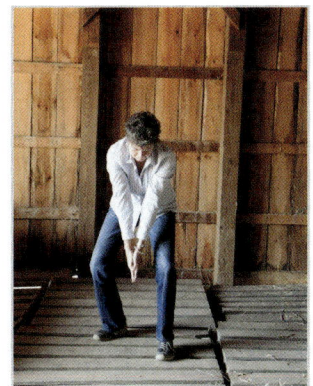

One notable feature of point shooting is the crouch position. Crouching under stress is an intuitive response. Your knees are bent. You may, or may not, use your gun's sights. Shoot Isoceles or Weaver up top, It's a quick, reflexive, stable shooting stance and method.

shooting and best practices was excellent. The environment on the range was professional, instructive, and safe. The first half day, we received classroom instruction. For the next day and a half, we shot 400 rounds at targets that were 20, 15, 10, and 5-feet away. The front sights of our guns were taped over.

Point shooting works like finger pointing. Your brain focuses where you point your index finger. This natural movement translates into instinctive alignment with your target. With proper instruction, point shooting can work as well with one hand as it does with two. Lots of tattered torso targets demonstrated the efficacy of this method.

The good news: point shooting, being an instinctive style, is quickly grasped by most shooters. Your Dad or Granddad may have learned this same technique over a mere 2-3 hours before being shipped off to World War II.

A CRAZY BEND IN LIFE'S ROAD

From plinker to home defender is quite a bend in my life's journey. I'm fine with it and hope to never to have to use my firearm tools. Ask any of my friends and they'll tell you that I'm way into the Pursuit of Happiness. I just never thought I'd have to defend it. But I've studied my options, I'll do my best, and I'm not looking back. Many people prepare only for peaceful coexistence and abundance. Why not embrace both? I suggest that a truly balanced life encompasses Yin & Yang. Dove & Hawk, Sheep & Wolf.

MY PERSONAL MANTRA
Live Life Fully.
Follow My Intuition.
Intend for the Best.
Prepare for the Worst. And, Have a Plan.

LETHAL DEFENSE:
Concepts You Need to Know

WITHOUT THE HYPE, IT'S COMPLICATED If one could remove the emotions, politics, and media bias, an individual's decision to use a gun for home and self defense is still complicated. Consider the following:

The intent of using a gun for self defense in the home, in most states, is to stop an intruder that threatens assault, violence, rape or kidnapping. The important word here: **stop.**

In classes across the country, civilian home defenders are taught that to stop a bad guy is to shoot them in 'the center of mass'—the area of vital organs located in the center of a person's torso—a good-sized target. To "stop" does not imply to kill, nor does it imply to wound. To use a gun for home and self defense, means to point it at a human being, pull the trigger, have enough control in a high-stress second to make well-placed shot(s), and stop or drop the bad guy. If he's been shot and retreats, let him. **Don't** shoot a retreating person in the back.

You may be thinking along these lines, like I did: "I'm okay with using a gun for self defense, but I don't want to kill anybody! I'll shoot him in the knee. I'll hurt him, but not kill him". Well, consider that it may be all you can do to see in the dark, get to your gun, grip it, aim it, press the trigger AND place an effective shot.

Most people will never have practiced enough to hit a small, moving knee-sized target. If the intruder is high as a kite, he may not feel any penetrating bullets and still move toward you. It happens! As long as they continue to be an imminent, serious threat, continue to shoot.

PROTOCOLS FOR SELF DEFENSE SHOOTING

If you're going to use a gun for home defense, do it effectively:

1. Assume that a shooting 'event' will happen very quickly, within 4-7 seconds. Figure out ahead of time how you're likely to respond to different scenarios.

2. Access a firearm instantly; load it quickly. Or keep it loaded. Or keep it loaded and locked up. (Practice accessing your gun in the dark). Seconds matter.

3. Prepare to shoot from a safe place, ideally one with cover to provide some protection in case the bad guy's armed. Minimize your exposure. Have steady support for your hands and a clear line of sight.

4. Know what you're aiming at. (Need eyeglasses? Keep a dedicated pair in that spot). If it's dark in your house, place a motion-activated light to backlight the bad guy and not shine on you. Do this well in advance. Like, now?

5. Work out a protocol to be damn well sure that the "intruder" is not your partner coming in late, or your teenager sneaking back into the house.

6. Make sure **no one** is behind the wall where you're aiming to shoot. You don't want your rounds to hit a family member or adjacent apartment dweller. You're legally responsible for where each bullet ends up.

7. If time allows, call police. With the phone open to 9-1-1, loudly tell the intruder to leave, state that you're armed, and prepared to shoot. Don't hang up. Policemen on the way (to this crime in progress) are listening. Inform the bad guy you called the police, and they're on the way. Don't hang up. Let your call be recorded by the 9-1-1 dispatcher. (Minimize talking, focus solely on the intruder).

8. Only use ammunition appropriate for shooting inside the home. Use frangible or expansion defense rounds that expand upon impact inside their target. Do not use rounds that are demonstrated to overpenetrate your target. If you miss, you don't want rounds to travel through interior or exterior walls. Research your ammo.

9. Under extreme stress, expect adrenaline to flow (See pages 106-107). Your hands may or may not be shaking. Breathe. Steel yourself. Find internal strength.

10. If the threat is real, if you've done your best to retreat (over a dozen states require this), if you can justify shooting to stop an imminent violent threat, then shoot for the center of mass. Even if you're an accomplished shooter, shots to the pelvis, heart, or head may be difficult to execute. Aim for the torso high chest.

11. Shoot until the threat stops, or he retreats from your property.

12. It's unlikely that 1 shot will stop a bad guy. Be prepared to fire 2, 3, 4, 5 rounds to stop the threat. But emptying 15 rounds into a bad guy won't serve you well (in court) when 4 or 5 rendered him harmless and bleeding out.

13. Stay vigilant. Stay armed. Reload if necessary.

14. Call the police now, if you hadn't done so before. Briefly, succintly state there's been a shooting. State that you're the homeowner. Describe what you look like, and what he looks like. Call your defense attorney right now or as soon as you possibly can.

15. Don't take your eyes off the bad guy until police arrive.

16. When police arrive, do 3 things:

    1) let them in;

    2) identify yourself and point to the bad guy;

    3) do not describe the event to police. If it's true, say only that "I was afraid for my life". Anything else you say, **can and will be held** against you.

17. Call your defense attorney now, if you didn't before.

It's the job of the police to apprehend the bad guy, take a report, and call an ambulance or coroner if needed. It's your job not to incriminate yourself. Let your attorney handle it.

!

CAUTION
Ignore what you see on TV!

Never go looking for an intruder.

Don't go down the stairs to find a bad guy.

Assertively demand a bad guy leave immediately.

Never enter into a conversation with them.

AMMUNITION &
THE HUMAN BODY

Most gun owners will never have to use their gun for self defense. *Yay! Hallelujah!* But if you intend to use firearms for home defense (and conceal carry), have a sense of what bullets do to people. Hopefully, you become a better shooter for it. Below, a short list of probabilities, facts, causes and effects:

- It's highly likely that a home invader will retreat upon the sight of a gun; and, will likely retreat after having been shot at, or shot once.[7]

- When someone is hit with a handgun projectile, whether a 9mm or .40, it's unlikely they'll fall backward, or fall down. It's more likely they'll keep moving until they pass out from blood loss, or their blood pressure drops and deprives the brain of oxygen. A true kill shot that severs the spine will drop a person dead. (Very rare).

- To be hit with a 9mm projectile is the equivalent of 10 pounds of force dropped on your chest from a height of 0.75 inches.[8] I've watched (too many) videos that show hits to a person's body, where they continue to stand erect or move around, slowing down as they bleed out.

- A bad guy who's enraged, psychologically imbalanced, and/or high may not feel a gunshot wound and still move toward you. This is very dangerous and can warrant multiple shots to stop them.

- Ammunition design comes in three general categories with different effects in the human body: penetration; expansion (on impact); or fragmentation (on impact). Full metal jackets (FMJ) penetrate. Hollow points expand. Frangible rounds fragment.

- Expansion rounds can create significant internal damage. These projectiles go from sleek-looking rounds to shredding machines upon hitting resistance and opening up. They're designed to create quick and extensive damage to blood vessels, major veins and/or organs. Heavy clothing, like denim, however, can clog the bullet's tip and prevent proper expansion. The bullet may still penetrate, but with less than intended damage.

- Small caliber ammunition, like .380s and .410s tend to penetrate, not expand, inside a body. There are exceptions in a .410 caliber, like the Winchester PDX1, it expands via multiple projectiles.

- If you research your ammo (a good idea!), stick with the FBI's recommendations for penetration into the human body (simulated using ballistic gelatin). Bullets that consistently penetrate 12" to 18" are considered ideal.

- There are a lot of reasons why a bullet that consistently averages 18" into ballistic gel, will not exit a thin person, for example. Chalk it up to: clothing, tensile strength of skin, muscle mass, muscle density, bones, angle of entry, and weird deflections if the bullet ricochets internally.

- Unlike ballistic testing gel, every human body is different. No one can predict exactly what a bullet will do inside a person's body.

Don't call this a bullet. Call it a "round" or a "cartridge". A single round of ammunition is made up of 4 parts: cartridge case, propellant, primer, and a bullet.
A "bullet" is the piece of lead that hits the target.

- Shooting someone in the center of mass may produce little physical effect, a bullet can becomed lodged in a body with minor disruption. A well-placed shot can take seconds for damage (bleeding) to show. Within those seconds, you may still be in imminent danger. Nearly all self defense instruction recommends shooting more than once.

- Stay away from boutique ammo designers who promote 24" or 30" penetration. It's inappropriate for home defense or conceal carry.

- Shotgun wounds cover a large surface area and lodge many BBs under the skin. The longer barrel of a shotgun (18.5" minimum) helps increases the likelihood of shooting what's aimed at. These reasons and more make shotguns a premier home defense choice. A shotgun slug, (a sizeable ball of lead) can create a major wound cavity that greatly disrupts internal organs, due to its diameter, mass, and velocity. But leave slugs to policemen who need to penetrate car doors. The likelihood of over penetration inside a home makes them unsuitable.

- If you do shoot someone, you're more likely to wound them than kill them. And you may not know if your last shot made a difference. Blood is unlikely to spew forth immediately; they're unlikely to fall backwards. Every situation is different. This is why multiple shots are taught in home defense when milliseconds matter.

- No self defense shooting will ever be like target practice. Or Hollywood.

GUN SAFETY

GUNS IN LOCKED CABINETS

Yes, one should keep all guns locked up in a very safe place that only 1-2 responsible people ever know about, or have access to. We've all heard the horror stories, kids finding and playing with Dad's loaded gun. Showing off his find, Sam hands his best friend Dad's loaded handgun. His friend playfully points it at Sam, and pulls the trigger. Little Sam never sees another day—shot unintentionally by his best friend who is now traumatized for life. Locked gun cabinets or gun safes are considered essential for owners of multiple guns. **Especially in households with children.**

GUNS IN OTHER PLACES

In preparedness for home defense, some people have guns in specific, concealed, never-disclosed-to-anyone locations. Others wear a holstered firearm; some sit in their favorite chair, gun placed down in the cushion. I knew a handicapped bachelor who kept a loaded revolver on his kitchen table. (It gave me the creeps, but that was the best solution for him). It's an individual choice depending on circumstances. Home defenders who live with other people, must be extremely aware, safe, and strategic in the placement of guns for home defense.

GUNS AND YOUR FAMILY

It is my opinion that you should tell members of your family that there is a firearm in the home ONLY IF they are sober, stable, and mentally well-adjusted older teens and adult(s) who you trust with your life, and, who've been well taught in the safe handling of firearms.

In an emergency, you need your family members to be on board and be part of the Response Plan. Each person needs to know and have practiced exactly what to do—retreat to a safe room; gather together in one defendable room; or hit the floor of the room they're in if they're not the armed defender—whatever works best given the nature of your family and layout of your home. Practice together. If firearms comprise any part of your plan, then everyone (over 14 or 17?) should have solid firearms training and a healthy respect of this potentially lethal tool.

GUN SAFETY FOR CHILDREN

Since 1988, the NRA's Eddie Eagle GunSafe® Program has reached over 26 million children in 50 states. This program was developed through the combined professional efforts of clinical psychologists, reading specialists, teachers, curriculum specialists, urban housing safety officials, and law enforcement personnel. Eddie Eagle's message stands to this day:

If you see a gun:

STOP!

Don't Touch.

Leave the Area.

Tell an Adult.

FAMILY DAYS AT LOCAL GUN CLUBS

The nightly news never seems to cover the Family Days held at the hundreds of small, respectable gun clubs across the country. These events welcome whole families to alcohol-free, picnic events that offer opportunities for the young and old to safely handle guns in the company of certified instructors. These social affairs give pre-teens, moms, dads, and grandmas a chance to learn and practice target shooting with range masters and instructors there to share knowledge and instruction in safe gun handling and target shooting.

(!) **NEVER EVER** place a loaded gun where a little kid or toddler can ever come across it. Not behind the TV, under the couch, or in a dresser drawer.

Don't be a moron about this!

—Calamity May

SHOOT & LEARN
Firearms I Own

Here I'm shooting Dr. Revolver's .44 magnum. Note the smoke (muzzle flash) and recoil. Hefty recoil like that is why one needs to practice a lot, and learn to bring the gun back down quickly, get on target, and shoot again.

I'm fortunate for the mentors who've taken time to train me during the course of this book. (And a big nod to all the guys at Wednesday night indoor range). I've shot dozens of semiautos, revolvers, shotguns, carbines, and rifles. What I do most often, however, is go to the gun club alone and practice. My first year, I put 1,500 rounds through my humble 9mm semiauto and 250 shells and slugs through my Dad's .12 gauge shotgun. My steadfast company consists of blue man targets. I like their center of mass. I know enough now to be dangerous. (And, yes, that is a well-intended pun).

I'm going to briefy describe three guns I have and lessons learned along the way.

SMITH & WESSON SIGMA 9VE

My friend, Terry, found a Smith & Wesson Sigma 9ve semiautomatic handgun for me. For $250, it came in a dry ammo case with 250 rounds, three 15-round magazines, and a holster—a pretty good deal I was told. Terry checked out its insides and mechanics and signed off. Knowing zilch about guns, I bought it.

New or used (if well maintained) Smith & Wessons are historically reliable. This low-end model isn't pretty or fussy; it is said to be durable.

This model has two notable features worth mentioning: an 10-pound trigger pull and no safety. You have to mean it to shoot it. At the time, being a new shooter owning one gun, it's what I learned on. Under the stress of adrenaline, the trigger pull will be a non-issue, and I won't have to remember to take off the safety.

LEARN BY DOING

The gun shoots well, except when I put cheap, light reloads in it. When I was a recent graduate of NRA's Basic Pistol class, I was solo at an outdoor range when it jammed. A brass casing failed to eject, it was half in, half out—stovepipe style. I'd been shown in class what to do, but now I was alone, inexperienced, and afraid of shooting myself! I entered into this monolog: "C'mon Suz, you can do this. You gotta. You can't put it in the car like this. You can't set it down Anywhere! And where the hell is the next round that's supposed to be up there?" I got a firm grip, physically and mentally, and turned the gun upside down. I pulled the slide back, shook it, and the casing fell out. I let the slide re-set. The next round was chambered. Problem solved!

My revolver mentor, whom I call Dr. Revolver, is an MIT-trained engineer. He's shot once a week since forever, and often shoots only one round at that. ("You've got to make the first shot count, Suzy.") He explained why the casing didn't eject: "Semi automatics need to have ammunition of sufficient weight and energy to cycle properly through the gun". Translation: The Sigma requires a heavier weighted round to eject spent casings and avoid misfires. The ammunition (reloads) I used were too light for that gun.

Dr. Revolver's .357 revolver and my Smith & Wesson 9mm, my Sigma 9ve, to the right. Note the difference in sizes of the guns and the rounds they fire.

LESSON LEARNED

If you're shooting semi-automomatic handguns, use reputable ammunition. Use only the ammo that's stamped on your gun's barrel and recommended by the manufacturer. Read your gun's manual. Buy quality rounds from reputable manufacturers.

Learn to clean your gun. Do it after every range visit. You'll get a better feel for your gun and develop confidence in this important tool. If it's hard to take apart, the Sigma is, take it to a gun smith to clean it for $30 after shooting. Some semiautos are notoriously difficult to disassemble and clean. If that's going to keep you from practicing, or cleaning your gun (and it may), you have 4 options: don't buy that gun, trade it in, take it to a gunsmith, or figure it out (read the manual and/or watch YouTube videos).

Semi-automatics have many positive qualities and come in many configurations. Some are fussy (lean toward jams and misfires); many have excellent reputations.

DAD'S ITHACA .12-GAUGE SHOTGUN

My Dad had one shotgun he used for bringing home duck and geese. My brother Ab inherited it, and when he learned about my endeavor, he gave it to me. *Thanks for the awesome gesture, Bro!*

You can read until you're blue in the face about how great shotguns are for home defense–most of it's true. Here's why:

- Shotgun shells deliver a spray pattern that's likely to hit some part of a bad guy. Good placement creates a significant wound cavity.

- The barrel's length helps you make more accurate shot placement.

- Shot placement doesn't have to be as exacting as a handgun–helpful under the stress of adrenaline.

- The "You've been warned" sound of pumping a shotgun, may have a survival-minded criminal running out the door. You still have to be willing to fire it, if they don't.

FROM GOOSE HUNTING TO HOME DEFENSE

Dad's .12 gauge shotgun came to me with its original 28" long barrel, a great length for waterfowl. But for home defense. it's too long for indoor use. Fortunately, I was able to change out the barrel in this model. For $100, a local gunsmith sold me his last 18.5" barrel (the shortest legally allowed), and installed it. I now have two barrels for this gun. I think Dad's alright with this modification (and intended use).

As per my point shooting class, I practice at target distances of 5', 10', 15' and 20'. I use full torso targets. Visually, they help me wrap my head around self-defense shooting.

Shooting 00 buckshot at these short distances, creates massive entry holes. It's my opinion however that 00 and slugs are unnecessarily powerful at short range in a house. I've switched to #7 bird shot. I happen to care about the walls in my home, who's behind them, and stray bullets departing my premises.

Shotguns used for hunting are only allowed to chamber three shotshells. For home defense, this shotgun can chamber five shells in the magazine when a thin wooden dowel rod is removed. That's how I keep it.

It may look cool, but shooting from the hip is not recommended. It's not a stable position, has less control, resulting in reduced accuracy.

LESSON LEARNED

In my experience to date, this powerful shotgun will not eject spent shotshells for only one reason–if I don't rack it with enough force. I gotta rack it forcefully like my Dad did, and so I practice.

Red arrows point to 00 buckshot shot from the hip at 20 feet. A .12 gauge shotgun, (Ithaca Model 37) loaded with 00 buckshot, (9 pellets each at 0.33-inch diameter) is too powerful for home defense. In real life, that shot off to the left (green circle) that grazed the target, could have penetrated a plaster wall behind it. I use #7 birdshot from a .12 gauge. At distance of 20-25 feet or less, it will have a tight grouping and significant stopping power.

These 4 hits are #7 birdshot from same .12 gauge shotgun, shot from the shoulder at 20 feet. Shoulder versus hip shots are much more stable, meaning your shots will most likely be more effective. Hits like this in the center of mass will slow a bad guy quickly. Factors to make that happen: shot placement, stable stance, and 4-point contact with shotgun, appropriate shotshell loads, and close proximity to the target.

Well placed shots to the center of mass, from a shotgun takes practice. Learn to shoulder it properly, have a solid stance, and practice racking a pump gun and get it back on target quickly. Accuracy first, then add speed.

And/or practice with your semiauto. Get used to what you have.

SMITH & WESSON MODEL 13 REVOLVER

My first handgun was picked out for me. The shotgun, I inherited. But I wanted a gun that I chose to be an ally in a self defense moment. This gun would have to be easy to understand, simple to clean, a straight shooter and feel good in my hand. After five range sessions with my mentor, Dr. Revolver, I knew it would be a revolver.

Smith & Wesson sent a batch of Model 13s to the Albany, NY police deparment in 1974. The gun shows light wear on the bluing (finish) from going in and out of a holster. The inside of barrel showed very light wear. Handguns used in police academies can be found in some gun shops and are cost effective.

I considered two. The first was recommended by Steve Barron, my point shooting teacher, a Smith & Wesson Model 60 .38 snub nose revolver. He carries one daily. He suggested the all-steel over the lighter Model 642 Airweight version, but likes both. So I tried both. I discovered that shooting lightweight (alloy or polymer) guns with short barrels demands a surprising amount of skill. It may seem counter-intuitive, but generally speaking, recoil from light guns is greater than recoil from heavier guns. Much practice is needed to develop grip strength, technique, and ability bring it back on target, quickly. Also, short barrels (as in 2.5") lack the inherent accuracy of longer barrels. You pretty much have to be an ace.

My search continued. An hour north, I found a Smith & Wesson Model 64, a .38 special, used police carry. It fit my hand like a glove. The weight was good; balance superb; ergonomics perfect. But I didn't buy it, because I didn't how to assess the condition of a used gun. So I waited....

Four months later, at the Monroeville Gun Show, I found one similar, but better: a 1974 lightly used Smith & Wesson Model 13, .357 revolver. It came from a police training academy over 40 years ago. I could hardly put it down. Still, I couldn't determine if the inside of the barrel was worn or in good shape. I walked away. Luckily, my warrior mentor, was at the show. He was happy to investigate.

He was impressed with the gun's overall condition: the rifling inside the barrel showed little wear; its mechanics were fine; walnut grips, dinged up but sound. I got a resounding Yes! Off we went to find an ATM.

I enjoy ironing shirts, but love cleaning my revolver.

LESSON LEARNED

I shoot this handgun well, not just because I practice with it almost weekly, but for its qualities that I suggest new shooters look for in a handgun: minimum 4-inch barrel, reasonable weight (steel over polymer), excellent ergonomics in hand-to-grip, and index finger-to-trigger, moderate ease of trigger pull (not too light, nor too heavy), and manufacturer's reliability.

Before getting to the final part in this firearms section, Recommended Firearms for Home Defense, I ask you to consider 3 Basics if you're thinking about buying a firearm for home defense.

3 BASICS IF YOU'RE BUYING A GUN

BASIC #1 CALIBER SIZE

Non-elite shooters, like myself, can take advantage of recent advances in ballistics technology. Powerful ammunition has been designed and tested to create significant wound damage, and have a greater likelihood to **not** penetrate drywall or plaster.

Besides not blasting through walls (and possibly reducing collateral damage), advances in ammunition means you don't need macho rounds like the .45, .40, or .357 calibers to stop a bad guy anymore. Small caliber guns chambered in 9mm, .380 and .410 have big stopping power these days. That's a plus for anyone who is unable or unwilling to manage more powerful handguns.

IN THE FLESH

For decades, Jackie has trained dogs for canine police work and has worked with many police men and women from many countries. She lives alone on 12.5 acres, and takes their stories of crime and violence to heart. On a daily basis, she has defensive plans in play: the handguns in her home are immediately accessible, loaded, unlocked, and ready to shoot. One of her tactics if pursued, includes "the Fatal Funnel" where she'd retreat to her bedroom behind her bed with a gun in that spot. In her U-shaped kitchen, (one way in and out), she keeps a gun near the cookie jar. Outside, knives are hidden in diverse locations. And of course, her dogs are trained to reduce an aggressive and violent intruder to a torn and bloody heap. I was in the presence of a true Calamity May. As I was about to leave, Jackie handed me a long piece of braided sweetgrass, and said, "I just love the smell of sweetgrass, I braid these every Fall. I can't help it, I'm part Cherokee".

BASIC #2 HEAVIER HANDGUNS OVER LIGHTER GUNS

Gun buyer, think about this: to shoot any gun is to manage a small explosive force. The more weight a handgun has, the more it absorbs that force (recoil). And, the longer the barrel, the straighter that projectile will travel toward where you aimed it.

Let's say you're a new, inexperienced, female shooter. You want a handgun for home defense. You bravely walk into a gun store by yourself to start looking. A salesman sees you as a naive consumer and takes you to the pink, lightweight 'ladies' guns and talks them up. (*Been there!!*) This overly applied sales pitch doesn't mean a light gun will work best for you! If that happens, ask to also handle a Sig Sauer P226 in .40, and the Smith and Wesson Model 67 in .38, and a Glock 19 in 9mm. Yes, they're bigger and somewhat heavier. But try them, their track records are outstanding. Try to find these guns (or similar) at a rental range before you buy your first, or, second handgun.

From experience, I find that heavier handguns have 2 main advantages over compact, light ones.

- Heavier guns have less recoil (good thing). Heavy = steel. The more the better. Ironically, the take less toll on your shooting hand.

- Heavier guns can have you back on target faster than snappy, light guns that travel north after every shot. In self defense shooting, you need to get back on target quickly. (page 135, #11, 12).

Consider these qualities in a handgun for home defense, and more:

Ergonomics. Find a gun that fits your hand well, where the first knuckle of your index finger comfortably reaches the trigger. (Though you press the trigger with the pad of your index finger).

Caliber: 9mm is a good minimum.

Barrel length: A 4" barrel will have you shooting straighter.

Go with what feels good to you. All the above factors have to work in your favor, not the friend you're with.

Test drive. Seek out a gun store that lets you rent guns to shoot. Attend a Ladies Day shooting event at a local gun club. They're free. Try multiple guns. Experience for yourself which ones feel good in your hand and which ones you shoot well with. (Hopefully, it's the same gun!)

BASIC #3 SHOTGUNS: PUMPS & SEMIAUTO

For decades, gun writers have lauded the virtues of the shotgun for home defense. Why? Here are good reasons:

- They're easy to shoot, but proper technique is still needed.

- Long barreled guns shoot straighter than any handgun.

- Lightweight shotguns (unlike light polymer or alloy handguns), work well for home defense.

- Shotguns have 4-point contact (shoulder/cheek/trigger/forend), which provides good stability while holding, aiming, shooting.

- With appropriate ammo selection, lightweight shotguns can stop a bad guy and missed shots *may* not blast through your walls.

- Shotguns shoot a spray pattern of pellets, some of which are likely to hit your target (versus the single round from a handgun).

- At short distances inside the home, at 10', or 20', well-placed shotgun rounds that hit the intended target (bad guy) deliver immediate and signficant wound damage.

- Shotguns come in pump action and semiautomatic. People with limitations in arm or hand strength may fare well with a lightweight semiautomatic.

- If you have a pump shotgun, the racking sound may encourage the intruder to leave. *(I hear this said all the time!)* But it may not. You still have to willing to shoot it.

I am a proponent of shotguns for home defense. It's my first go to tool for armed home defense.

A gun club I belong to put on a free Ladies Day Shoot sponsored by Ruger, I was one of 10 instructors on the range to help new shooters handle guns. Ten different guns of varying calibers were available. (8 semiautos & 2 revolvers). I manned a whopping big revolver that intimidated every woman that stepped up for her turn. "Don't worry", I said, "this gun will probably deliver your best shooting today." Sure enough, they all shot it remarkably well. In the end, 15 out of 20 women liked that gun the best.

Run with the big dogs!

RECOMMENDED FIREARMS
For Newbies, for Home Defense

Beta readers of this book asked for my recommendation for guns for home defense. Not an easy question. Choices today amongst shotguns, handguns and ammunition is vast. Expert and pseudo-expert opinions are equally broad. I'm going to keep it simple...

I'm narrowing this discussion to new shooters who want a firearm for home defense that's light, easy to use, affordable, and can stop a bad guy. These are guns that have a proven track record and long history of success.

SHOTGUNS

Shotguns, aka long guns, have worked for pioneer women in sod houses to modern women in thin-walled homes. They operate on simple mechanics and are easy to use. For close-range, defensive purposes, you're bound to get some lead in the bad guy. But you still have to practice!

Here are my recommendations for guns and ammuntion for use inside the home.

DEFENSE AMMUNITION

Winchester PDX1, .410

100 rounds, $13.50, typ.

Shell length: 3"
Shot type: Bonded (*a good thing!*)
Shot weight: 1/4 ounce
Muzzle velocity: 750 feet per second
Effective range: 15-20 feet or less.
Unique features: 3 brass pellets, and 12 BBs designed for close range engagement.

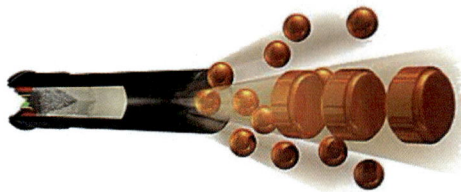

PUMP ACTION SHOTGUN

Mossberg .410 500 Tactical. Pump with Forend

Model #50359. MSRP $477

I like pump action shotguns. They're manually operated, their mechanics are simple, and they rarely misfire (unless you use junk ammo). This one's a short, light polymer shotgun with a steel barrel in a .410 bore. (Yes, polymer shotguns are a whole different story than polymer handguns). With it, you can hunt small game or stop a bad guy.

Paul Chandler, co-owner of Altra Firearms, recommends this shotgun for new shooters for home defense. His wife keeps one by her side of the bed. My girlfriend and I went to try it out. We both liked the forend grip better than its sibling with the

pistol grip. You pump (or 'rack') it by pulling back the forend and pushing it forward (to load a shell in the chamber). Yes, it has that Bad Guy Beware racking sound that only pump actions have. It's a smooth, easy action that's improved with decent gun oil lubrication when you take it out of the box.

At 5.5 pounds, it's light, (a tad light for my taste, but my girlfriend likes it). The barrel length at 18.5", the overall length at 37.25" makes it nimble. And the recoil? Minimal. NOTE: a .410 is a low-powered shotgun. To stop a bad guy, using a home defense load like the PDX1 (left) is mandatory.

Home Defense Ammunition: Winchester PDX1, .410, 2-1/2" shotshells

NOTE

Racking a pump action shotgun has a pre-emptive "You've been warned", will give away your position. Think it through.

ZONE 3 – SELF DEFENSE CATALOG

SEMI-AUTOMATIC SHOTGUN

Remington 11-87 Sportsman, Compact, 20 gauge, Semi-Automatic

Model #83626. MSRP $704.

Semi-automatic shotguns don't need to be pumped to chamber a shotshell. They automatically chamber a shell after every shot. That's a good thing for those who can't rack a shotgun due to a physical handicap.

Determine if your dedicated home defense shotgun is the right length for you.
Set the end of the stock into the inside bend of your elbow, with the gun laying flat on your forearm. Does the pad of your index finger easily reach the trigger? (This shotgun fits my arm length well). If it doesn't, have the stock cut down to the correct length by a gunsmith.

Like the .410, this compact .20 gauge shotgun is for people of smaller stature, but, anyone (even 6'2" stocky guys) can shoot it. This model has a 21" barrel; its overall length is 41.25". It weighs 6.5 pounds, and has a 4 + 1 shotshell capacity. (4 shotshells in the magazine, 1 in the chamber). With a gas-operated action, it's said to have very light-recoil. It lacks the (in)famous warning sound of a pump action shotgun though, but it's more important to have a gun that gets the job done. Automatic (versus manual) firearms need ammuntion of manufactured specified loads to cycle properly. In other words, don't use ammo that's too light, or too 'hot' for your gun.

The recently re-introduced 11-87 LOP has an adjustable length of pull. A .20 gauge shot gun is smaller than a .12 gauge; bigger than a .410. Commonly used for taking turkey, or rabbits, inside a home, with proper ammunition, it's a potent weapon. For those on a budget, used guns can be very reliable and affordable.

Home Defense Ammunition: Remington .20. 2-3/4" shotshells, 1 oz. #6 or #7 birdshot for distances of 25-feet or less.

THE SHOTGUN YOU INHERITED., OR, THE ONE IN THE GUN SAFE

Shoot What You Have!

Maybe it was a gift, an inheritance, or your partner's...

I inherited my Dad's Ithaca Model 37 shotgun that he used for hunting geese and duck. It's a heavy thing, its weight helps to absorb the significant recoil inherent to every shot. Number 7 bird shot isn't too bad. It's my defense load of choice in our small home. When practicing with slugs, it's a bruiser. (Note: in my opinion, slugs have no place in defense inside a home, small or large. If you miss, they'll blast through walls interior walls (and flimsy exterior ones), And, like a handgun, a slug is one projectile—will you hit your target?).

Dad's gun came with a 27" long barrel. I was able to swap it out for an 18.5" barrel—a much more manageable size for indoor use.

If you inherited, or need to learn on a shotgun already in the house, first have it professionally cleaned. Then learn to handle it, load it, unload it and shoot it safely. Learn from a certified instructor, or experienced hunter you trust. For me, at first, I had problems with spent shells not ejecting properly, it wasn't the gun's fault. I didn't rack it hard enough. I've learned to rack it like my Dad, with speed and force. After shooting, clean your shotgun, it's easy and satisfying.

Modifications I made to the Model 37 for home defense:

- installed an interchangeable 18.5" barrel on it

- put a LimbSaver recoil pad on the butt stock

- made it sling ready

- checked for length of pull (it was fine, already reduced by a 1/2-inch).

- put a shot shell band on the stock.

Home Defense Ammunition: Remington .12,. 2-3/4" shotshells, 1 oz. #7, at distances of 25' or less.

SHOT SHELL SIZES

SHOTSHELLS & HUNTING

Shotguns and shotshells come from the hunting world. Shotshells are filled with small, uniformly sized pellets called 'shot', typically steel. Buckshot is significantly larger than shot.

Shotshell loads for hunting game are:
Rabbits, #7.
Ducks, #5, 6
Deer, 00 (pronounced "Double ott buck").

SHOTSHELLS & HOME DEFENSE

Shotguns are powerful for home defense because distances are short, (typically 10 to 25 feet); shot patterns stay tight, and velocity is good for the task. Combined with good shot placement, these factors can make for damaging wound cavities to stop a bad guy.

Smaller caliber shotguns are well suited for home defense, like the .410 bore, .20-, or .16-gauge shotgun. But if 12-gauge is what you have, use it!

A shotshell with 1 ounce of shot contains:

317 steel pellets in #6

429 steel pellets in #7

Penetration of pellets into a person is highly variable, depending on shot placement, clothing layers, and many other factors.

Penetration into drywall is highly variable depending on wall composition, shotgun size, shot size, shotshell loads, number shots fired.

Let me reiterate what online ballistics testing expert ShootingtheBull concludes[7.]

"Most bad guys will retreat at the sight of a gun. Those who continue toward you will retreat when shot at, or hit once. A very small percent will continue to move in. If they do, stand your ground."

IMPORTANT MISCELLANEOUS

- Never go looking for an intruder. Don't go down the stairs to find a bad guy. *Ignore what you see on TV!*

- Except to loudly and assertively tell a bad guy to leave (once is enough), never succumb to a conversation with them, even a teenager. *Ignore what you see on TV!*

- Don't assume a downed intruder is dead. Assume they're playing possum. They may not stay that way. Keep your distance. Be vigilant and ready.

- Don't shoot a downed intruder 26 times because you're angry. You'd probably go to jail for excessive use of force, or face a stiff civil suit penalty.

- Always take a stand from your strongest position. Strategically and legally.

What you practice is what you'll do under stress.

Have a Plan and Backup Plans.

Practice.

Make sure your gear works, and be able to reach it in the dark.

BRIEF INTRODUCTION TO SELF-DEFENSE LAWS*

KNOW YOUR STATE'S SELF-DEFENSE LAWS

Ignorance
of the law
is not
a legal
excuse!

—U.S. Law Shield

This last topic very briefly touches upon self defense law in the United States. The reasons for brevity: 1) the topic is well explained by self defense attorney/authors to whom I refer you; and, 2) whether you wield a frying pan or a firearm, you need to understand different legal aspects of self defense, such as context, threat discernment, and use of force as defined in your state.

As a home defender, the burden and responsibility of self defense is upon you. Be well informed. Know your legal ground. Develop your plans and backup plans accordingly. If potentially lethal weapons are part of your plan, find an excellent criminal defense attorney who's knowledgeable of the self defense laws in your state. Get their business card before you need it.

The fight to save your life or a family member's life is the first one. The second fight could be a legal one. The more you know about your state's laws and act accordingly, the greater your chances of not going to jail or losing all that you were intent on protecting.

VOCABULARY

I present legal words in the likely sequence of a potential event. Abbreviated definitions are taken from (or shortened from) Cornell's Legal Information Institute (www.law.cornell.edu), unless otherwise noted.

REASONABLE PERSON A phrase used in tort and Criminal Law to denote a hypothetical person in society who exercises average care, skill, and judgment in conduct and who serves as a comparative standard for determining liability. [legal-dictionary.thefreedictionary.com]

The hypothetical reasonable person behaves in a way that is legally appropriate. Those who do not behave at least as a reasonable person would, are considered negligent and may be held liable for damages caused by their actions.

This applies to all self-defense means and tools covered in this book.

REASONABLE BELIEF When there exists a reasonable basis to believe that a crime is being or has already been committed.

CASTLE DOCTRINE (AKA STAND YOUR GROUND) The Castle Doctrine is **an exception** to a rule in place in some jurisdictions that requires a defendant to retreat (Duty to Retreat) before using deadly force in self-defense. The Castle exception states that if a defendant is in his home, he is **not** required to retreat prior to using deadly force in self defense.

IMMINENT Generally, imminence is time-based. Massad Ayoob, (an internationally recognized author, firearms instructor, expert witness), developed this three-prong approach to help determine if a threat is imminent:

> **Ability.** The person threatening you or others has to have the ability to kill or cripple. (Is my attacker able to hurt me?)

> **Opportunity.** The attacker has to have the opportunity to use deadly force. (Can their attack get to me?)

> **Jeopardy.** The bad guy has to manifest his intent to harm by words or actions that would be interpreted by a reasonable and prudent person as putting you in imminent danger of death or serious bodily injury. [*The Law of Self Defense*, Andrew Branca]

NECESSARY Use of force is justified when a person reasonably believes that it is necessary for the defense of oneself or another against the immediate use of unlawful force. However, a person must use no more force than appears reasonably necessary in the circumstances. [Lectric Law Library].

SELF DEFENSE The use of force to protect oneself from an attempted injury by another.

SERIOUS BODILY INJURY Injury that is life threatening, causes permanent disfigurement, or results in permanent disability. [*Self Defense Laws of All 50 States*, Mitch Vilos and Evan Vilos.]

DEADLY FORCE Legally speaking, deadly force is any force that can cause death, permanent disfigurement, serious physical injury, long-term damage to the function of any part of the body, or rape. Non-deadly force is any force that doesn't do any of those things. [*The Law of Self Defense*, Andrew Branca.]

> There are 51 totally separate legal systems in the United States. Each state has its own independent court system, its own constitution, and its own laws.
>
> —Karen L. MacNutt

4 EXAMPLES OF SELF DEFENSE LAWS: PENNSYLVANIA

Here are a few nuggets of self defense laws in my state, Pennsylvania. They relate only to self defense in ONE'S HOME. I drew from two highly useful and complementary books both published in 2013, both written by attorneys specializing in self-defense law: *The Law of Self Defense* by Andrew Branca, and, *Self Defense Laws of All 50 States* by Mitch Vilos and Evan Vilos.

Duty To Retreat (where one must withdraw or retreat to a safe place prior to using force even if you are the non-aggressor and threatened by the bad guy): Pennsylvania: No Duty to Retreat in one's home. (As of June 28, 2011).

Other states and territories:
• 30 states have No Duty to Retreat;
• 13 states say that a person has the Duty to Retreat "only before using deadly force"
• 4 states have Duty to Retreat before using any force.
• 2 states have specific conditions.
Note: if a homeowner, anywhere, can prove retreat prior to using deadly force, it's typically well regarded should your case end up in front of a jury.

When Deadly Force is Justified....Specific Acts of Violence
Pennsylvania: A defender must believe that such force is necessary to protect herself or himself "against death, serious bodily harm, kidnapping or sexual intercourse compelled by force or threat".

    Exception 1: If that other person legally lives in the same dwelling as you, such as a roommate or spouse. HOWEVER, if that roommate or spouse pulls a gun or knife on you or goes to choke you, you are entitled to use deadly force.

    Exception 2: Your legal defense may go south IF you are involved in any illegal activity; and/or IF you provoked the use of force against you in that same encounter; and/or IF you illegally possess firearms, etc.

When Deadly Force is Justified....Forcible Entry
Pennsylvania: "to terminate unlawful entry into defender's dwelling"; and, "to prevent the commission of a felony in a dwelling."

When Non-deadly Force is Justified....
Pennsylvania: "when the defender believes that force is immediately necessary for self protection against another person using unlawful force in that moment".

Use of Deadly Force in Defense of Property:
Pennsylvania: Allowed, with "limitations on justifiable use of force", they are:
- Unlawful entry into one's dwelling;
- The homeowner (actor) has reason to believe that nothing less than deadly force would terminate the entry;

If you're a Pennsylvania resident intending to use defense tools or firearms for home defense, you must read Pennslyvania code, P.A.C.S.A. Section 505 (2) through (2.6). I especially liked Vilos' book for explicit breakdowns.

This bears repeating: you need a good defense attorney with a strong suit in self defense laws in your state. Find one before you need one; consult with them to clarify legal issues for your situation. Have their business card or phone number at hand in the case of a fateful event.

It's critically important that only your attorney handles questions posed by police and authorities following an event. When the police say "everything you say can and will be held against you," they mean it. Don't incriminate yourself.

BOTTOM LINE
HAVE A PLAN, AND BACK UP PLANS!

EPILOGUE

Learning to be at home in the real world was uncomfortable at first. I began by putting one foot in front of the other, very awkwardly at times, and worked on all the things covered in this book: environmental awareness, intuition, core strength, clean diet, self-defense classes and firearms training. I felt different, more grounded, calmer. The rough fact that I'm responsible for my personal safety and the safety of my home began to smooth out and feel right. Yes, it takes time. A Queen of Denial is not unmade in a day. Cultivating and enhancing one's personal power, in all its aspects, is the work of a lifetime. Personal evolution is why, I believe, we're here.

Having embarked on this journey and been under sail for several years, I give it a 5-star rating. I'm a better person for it. I can only encourage you to cultivate or enhance your personal power. Implement just one aspect of this book. Install a good deadbolt. Let it sink in. Try another tool. Build upon small successes. (Success=feeling good or confident). You're the only one who can tap into the power of your intuition; or experience with heightened awareness the world around you. Feel Primal You wake up. Strengthen your home and sleep well. Know what you're willing to do. That alone affects all aspects of your life. Take up tai chi or close quarters combat, or both. Learn to safely shoot a gun and hit your marks. If you're only slightly open minded—try it. That alone is an accomplishment!

The world doesn't need more guns, predation or domination. It needs more people who embrace personal responsibility and are willing to act accordingly—with grace and mettle. A greater good is renewed when individuals choose to be responsible for their own safety, health, and happiness. I look forward to helping people develop strengths and reduce fears. It's another step forward in gaining solid footing in an unstable world.

Thanks for getting this far. Let's kick it up a notch by sharing stories, solutions, and insights. Find me on calamitymay.com. Email me, or comment on a blog post. I look forward to saying hello and hearing from you.

I'd be happy to speak to your organization, put on a shotgun seminar, and/or have a book signing near you.

May all your calamities be averted!

Suzy Meyer

September 20, 2017

REFERENCES

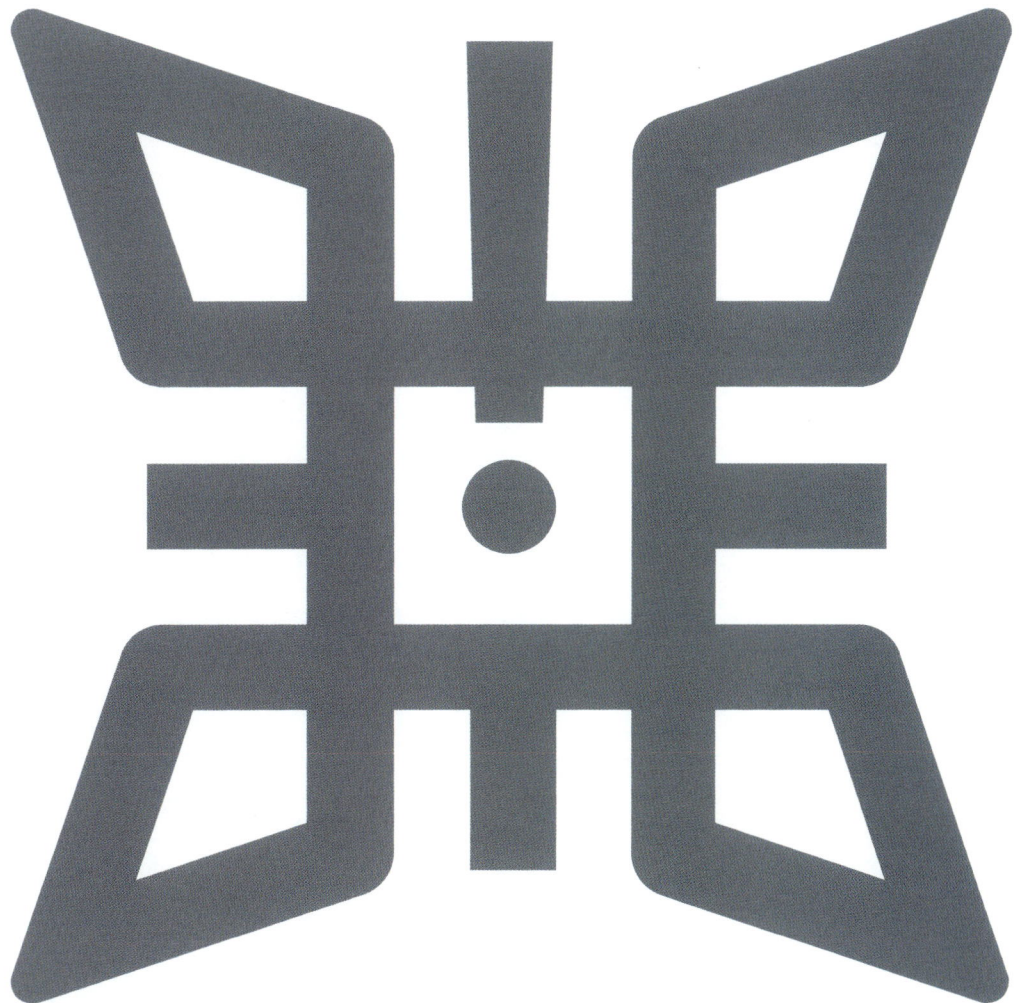

APPENDIX A

FBI Definition of 8 Crimes Tallied on the Uniform Crime Report (UCR)

The UCR Program collects statistics on offenses known to law enforcement. In the traditional Summary Reporting System (SRS), there are eight crimes, or Part I offenses, that are reported in the UCR Program. These offenses were chosen by the FBI because they are serious crimes that occur with regularity in all areas of the country; and, they're likely to be reported to police. FBI's definitions of Part I offenses are below:

Criminal homicide

a.) Murder and non-negligent manslaughter: the willful (non-negligent) killing of one human being by another. Deaths caused by negligence, attempts to kill, assaults to kill, suicides, and accidental deaths are excluded. The program classifies justifiable homicides separately and limits the definition to: (1) the killing of a felon by a law enforcement officer in the line of duty; or (2) the killing of a felon, during the commission of a felony, by a private citizen.

b.) Manslaughter by negligence: the killing of another person through gross negligence. Deaths of persons due to their own negligence, accidental deaths not resulting from gross negligence, and traffic fatalities are not included in the category Manslaughter by Negligence.

Forcible rape

As of December 2011, the UCR Program revised its SRS definition of rape: "Penetration, no matter how slight, of the vagina or anus with any body part or object, or oral penetration by a sex organ of another person, without the consent of the victim." The effect of this definition change will not be seen in reported crime data until after January 2013. Data reported from prior years will not be revised.

Former definition: The carnal knowledge of a female forcibly and against her will. Rapes by force and attempts or assaults to rape, regardless of the age of the victim, are included. Statutory offenses (no force used—victim under age of consent) are excluded.

Robbery

The taking or attempting to take anything of value from the care, custody, or control of a person or persons by force or threat of force or violence and/or by putting the victim in fear.

Aggravated assault

An unlawful attack by one person upon another for the purpose of inflicting severe or aggravated bodily injury. This type of assault usually is accompanied by the use of a weapon or by means likely to produce death or great bodily harm. Simple assaults are excluded.

Burglary (breaking or entering)

The unlawful entry of a structure to commit a felony or a theft. Attempted forcible entry is included.

Larceny-theft (except motor vehicle theft)

The unlawful taking, carrying, leading, or riding away of property from the possession or constructive possession of another. Examples are thefts of bicycles, motor vehicle parts and accessories, shoplifting, pocketpicking, or the stealing of any property or article that is not taken by force and violence or by fraud. Attempted larcenies are included. Embezzlement, confidence games, forgery, check fraud, etc., are excluded.

Motor vehicle theft

The theft or attempted theft of a motor vehicle. A motor vehicle is self-propelled and runs on land surface and not on rails. Motorboats, construction equipment, airplanes, and farming equipment are specifically excluded from this category.

Arson

Any willful or malicious burning or attempt to burn, with or without intent to defraud, a dwelling house, public building, motor vehicle or aircraft, personal property of another, etc. Arson statistics are not included in this table-building tool.

Source: U.S. Department of Justice, FBI, Uniform Crime Reporting Statistics web site: http://www.ucrdatatool.gov/offenses.cfm

APPENDIX B

More Newspaper Headlines that Demonstrate Underreporting Crimes or Downgrading Crimes

Times Picayune, October, 2013
NOPD Crime Statistics May Be Underreported, Audit Finds
in 2012, "319 of 1,000 serious incidents went unreported to FBI....2 of 20 serious crimes reviewed by auditors were imporoperly downgraded".

California Colleges - University Herald October 11, 2013
USC, Occidental College Admit Clery Act and Title IX Violation in Underreporting Sex Crimes on Campus

PoliceOne.com, June 20, 2012
Fudge Factor: Cooking The Books On Crime Stats. What Do Police Report Writing and Fudge Have in Common?

LawReport.org, October 25, 2012
Denver Police Have Underreported Crime on Public Website Since 2009.

Bangor Daily News, February 20, 2004
Atlanta Underreported Crimes to Land Olympics, Report Says

APPENDIX C

U.S. Dept. of Education Releases List of Higher Education Institutions with Open Title IX Sexual Violence Investigations

Elementary, secondary and post-secondary schools (colleges, universities) that receive federal funds must, under the Jeanne Clery Act, report crimes and attempted crimes that happen on campuses and in on-campus housing. Under President Obama's administration, the first list of schools under investigation for possible crimes relating to sexual violence (rape), dating violence, domestic violence, and stalking was published.

On May 1, 2014, the U.S. Department of Education's Office of Civil Rights explained the process in a press release. The list of schools includes "investigations opened because of complaints received by the Department of Education's Office for Civil Rights (OCR) and those initiated by OCR as compliance reviews. Schools on the list below should not be construed to have actual mishandling of any Title IX-related incident until the results of that review is made public". See DOE web site for resolved investigations.

This list of schools in violation and those cleared of violations, is updated weekly. One can acquire a list of elementary through post-secondary schools with potential Title IX violations, for free, by emailing ocr@de.gov. Or calling, 800.421.3481. Or DE's press office, 202.401.1576.

The original press release (which lays out intent) can be found here at: www.ed.gov/news/press-releases/us-department-education-releases-list-higher-education-institutions-open-title-), or, use the date to search for it.

A PDF from January 2017 listing 225 schools is viewable at calamitymay.com (sadly or oddly, that list did not include elementary or secondary schools).

APPENDIX D

Myth 1 "The Police Will Protect Me"—Five Court Cases that Establish the Contrary

I've learned that it takes tragic, gut-wrenching cases to be heard in state supreme courts, and the Supreme Court of the United States. Through judicial process, and rigorous interpretation, opinions are delivered by judges and Justices, respectively, that underscore the enduring meaning of laws we live by. To believe that police or state services exist to 'protect' you is a dangerous misconception.

Below, Supreme Court law unpacked, and other prime examples listed.

U.S. Supreme Court

DeShaney v. Winnebago County Department of Social Services. 489 U.S. 189 1989.

This tragic case tested the Constitution's emphasis on government services (in this case, social services or police) in not interfering with the private acts of citizens. Joshua DeShaney was an infant when he came to be in his father's custody in Wisconsin. For four years, his father brutally beat Joshua while he was under the supervision of Winnebago Social Services. The child was left partially paralyzed and with permanent brain damage. His (estranged) mother sued the Winnebago County Dept. of Social Services for money to cover his high costs of care, that she claimed were caused by conditions stemming from a lack of protection on the part of child welfare services. The case went to the U.S. Supreme Court in 1988. Justice William Rehnquist penned the 6-3 opinion of the Court, that is not without compelling dissent. The majority, decided or upheld, that:

- The State has no duty to protect a citizen from the bad actions of another citizens [even an infant and his abusive father-SM];

- The government has no duty to safeguard people and protect them from privately inflicted harms.

- The State cannot be sued for negligence or failure to prevent harm or provide protection (DeShaney case cited).

Independent Interpretation of DeShaney

Attorney Lance LoRusso, former law enforcement officer and General

Counsel for the Georgia Fraternal Order of Police, helps us understand the Court's opinion by underscoring significant phrasing in Justice Rehnquist's opinion. Russo used italics to highlight meaning:

"Nothing in the language of the Due Process Clause itself *requires* the State to protect the life, liberty, and property of its citizens against invasion by private actors. The Clause is phrased as a *limitation* on the State's power to act, *not as a guarantee of certain minimal levels of safety and security*. It forbids the State itself to deprive individuals of life, liberty, or property without 'due process of law'…and was intended to prevent government "from abusing [its] power, or employing it as an instrument of oppression. *Its purpose was to protect the people from the State, not to ensure that the State protected them from each other*."

-from http://www.lawenforcementtoday.com/2012/08/26/active-shooters-and-the-role-of-law-enforcement-it%E2%80%99s-not-about-liability/

Warren v. District of Columbia, 444 A.2d 1-D.C. Ct. of Ap. 1981.

To describe the horrible events that befell three women roommates, I defer to this brief description from the web site, *What Really Happened*, www.whatreallyhappened.com/WRHARTICLES/courtrulingsonpoliceprotection.php:

"Two women were upstairs in a townhouse when they heard their roommate, a third woman, being attacked downstairs by intruders. They phoned the police several times and were assured that officers were on the way. After about 30 minutes, when their roommate's screams had stopped, they assumed the police had finally arrived.

When the two women went downstairs they saw that in fact the police never came, but the intruders were still there. As the Warren court graphically states in the opinion: "For the next fourteen hours the women were held captive, raped, robbed, beaten, forced to commit sexual acts upon each other, and made to submit to the sexual demands of their attackers."

The three women sued the District of Columbia for failing to protect them, but D.C.'s highest court exonerated the District and its police, saying that it is a "*fundamental principle of American law that a government and its agents are under no general duty to provide public services, such as police protection, to any individual citizen*."

Riss v. City of New York, 22 N.Y. 2d 579, 293. NYS2d 897, 240 N.E. 2d. 860 (N.Y. Ct. of Ap. 1958);

Linda Riss, an attractive young woman, was terrorized for six months by a rejected suitor, attorney Burton Pugach. Repeatedly, Pugach threatened to have Linda killed or maimed if she did not yield to him: "If I can't have you, no one else will have you, and when I get through with you, no one else will want you." In fear for her life, she went to police, to no avail. On June 14, 1959 she became engaged to another man. At a party celebrating the event, she received a phone call warning her that it was her 'last chance'. Utterly distraught, Linda called the police and begged for help. Protection was refused. The next day Pugach carried out his dire threats, and hired thugs who threw lye in her face. She was blinded in one eye, lost partial vision in the other; and her face was permanently scarred. After the assault the authorities concluded that there was some basis for Linda's fears, and for the next three and one-half years, she was given around-the-clock protection. Linda sued the City of New York that they be held liable in damages for its failure to protect her from harm. The court's decision: *"The government is not liable even for a grossly negligent failure to protect a victim of a crime."*

Judge Keating, wrote in his dissenting opinion, "What makes the City's position particularly difficult to understand is that, in conformity to the dictates of the law [nohandguns allowed for self defense in New York-SM], Linda did not carry any weapon for self-defense. Thus by a rather bitter irony she was required to rely for protection on the City of New York which now denies all responsibility to her."

Lynch v. N.C. Dept. of Justice, 376 S.E. 2nd 247 (N.C. App. 1989).

The North Carolina Supreme Court states: *"Law enforcement agencies and personnel have no duty to protect individuals from the criminal acts of others; instead their duty is to preserve the peace and arrest law breakers for the protection of the general public."*

APPENDIX E

Myth 2, Castle Rock v. Gonzales, narrative of the incident and legal

United States Supreme Court Town of Castle Rock v. Gonzales, 545 U.S. 748

This case shows us that police have discretionary powers to choose if and when they may pursue crimes of domestic abuse, even if that state's language calls for "mandatory" intervention. This has far-reaching impacts for women, men, and children subjected to domestic violence.

In 1999, in the town of Castle Rock, Colorado, Jessica Gonzales obtained a restraining order on her husband, Simon. On previous occasions at home, he threatened to kill his wife and three young girls; once, he tried to hang himself in front of his children. Mrs. Gonzales went before the Colorado Court and was awarded a temporary restraining order–Simon Gonzales was ordered to move out and stay 100 yards away from the home. Six weeks, later under a permanent restraining order, he was allowed visits at the home. It was then, that he illegally 'took' his three daughters who were playing outside in their front yard. Jessica contacted police and waited hours for them to arrive to so that she could give them the restraining order. She called seven times and went to the police station twice. When the police finally did show up, they deemed that since the girls were with their father that they would be "fine."

At 3:20am Simon Gonzales pulled up in his pickup truck in front of Castle Rock's police station and started shooting toward the station. Police returned fire, killed him, and riddled his truck with bullets. The three young girls were found dead in the back of the pickup. (Some accounts say the cab of the truck). It remains unresolved, whether they died from gunfire during the shootout; or whether their father shot them hours earlier as they slept in the back of the truck. (Colorado Bureau of Investigations did determine that the bullet casings found in the back of the truck were "of a different caliber than those issued by Mr. Gonzales' gun.")

Jessica Gonzales brought suit against the police department of Castle Rock, which went on to the United States Supreme Court in 2005. The Supreme Court decided in a 7-2 vote, that Jessica Gonzales

a) *has no right as a citizen to expect police enforcement of a restraining order* (despite Colorado's statutes for mandatory arrest of named violators on restraining orders); and,

b) *cannot expect police protection* (citing DeShaney v. Winnebago. See Myth 1); and,

c) *cannot sue police for lack of protection, or their failure to make arrests for crimes in progress.*

On June 28, 2006, The New York Times reported, "The Supreme Court ruled on Monday that the police did not have a constitutional duty to protect a person from harm, even a woman who had obtained a court-issued protective order against a violent husband making an arrest mandatory for a violation."

Jessica Lenahan (formerly Gonzales) went on to bring the first case brought by a domestic violence survivor against the U.S. government to an international tribunal, the Inter-American Commission for Human Rights (IACHR), the legal and administrative body of the Organization of American States. She presented the violations against her children and herself; and, the lack of remedy in the US justice system. Six years later, in 2011, the IACHR decided in her favor. The Commission recognized that the United States government allows human rights violations against women and children. The tribunal also set forth comprehensive recommendations for changes to U.S. law and policy pertaining to domestic violence. (The findings and recommendations of the IACHR however, has no force of law in the United States). Jessica's Lenahan's persistence and passion, however, has inspired and fueled advocates who work tirelessly on behalf of victims of domestic abuse.

APPENDIX F
Considerations for New Doors Made from Wood, Fiberglass, & Steel

Replacing Our Front Door

To replace our 100-year old wooden door, we first considered fiberglass for its low maintenance, attractive finishes, and insulation values. In the end, we chose wood for three reasons:

- impact resistance

- ability to be shaped and custom sized at a minimal cost (to fit our non-standard doorway) and

- authentic wood grain and color

During installation, our carpenter noticed the bowed lintel above the interior front door. (It is a 102-year old bungalow). Having a wood door gave us the option to have it planned on site to fit this deformation. You can't do this with fiberglass or steel.

Consumer on a Quest

As a consumer setting out to research and buy a new front door, I found myself entering a market that sells to homeowners based solely on aesthetics. What about specifications for security? I learned that there are no measures to gage a door's security. Two of the biggest wood door manufacturers associations rate their doors on quality and construction, nothing else. To be fair, much relies on the quality of the jamb and frame into which a door is attached—if the manufacturer doesn't supply the frame, they don't touch security issues.

This I did learn: Be wary of cheap doors. Buy the very best you can afford. Buy well-crafted doors from companies that can demonstrate quality materials, inside and out, and that use the best possible construction techniques.

Below, I highlight a few qualities of wood, fiberglass and steel doors.

Wood Doors

It is rare today to find solid wooden doors. I didn't know that. Solid wood doors in old building that are still in place 150 or 400 years later were constructed from dense premium hardwoods, like white oak. Today, that kind of door would cost a king's ransom. Back then, strong-grained woods were grown in managed woodlots to serve these purposes—it was the best possible material at the time. Less premium woods warp over time. Fabricating these old doors called for solid metal rods or bolts, and/or sturdy wood dowels made of yew (or other specific wood) to craft each door into a strong whole. Only battering rams broke through. They're not made like that anymore.

Today's two most common types of exterior doors are either paneled or flushed. On both, the outside edges (the frame) are made with wood. The door's interior is constructed with "structurally engineered" wood–a composite. They get their wood look from thinly sliced wood veneers, (1/16" to 1/4" thick), applied to the outside. All exterior wood doors must be stained or repainted every few years. The greater their exposure to direct sun and moisture, the more attention they need to keep the veneer in good condition.

Panel wood doors are popular for wide variations of patterns, from traditional to wildly contemporary. A paneled door is a framework of horizontal (rails) and vertical members (stiles) that are solid wood, typically 1-3/4" thick. They are joined with dowels, or, on better doors, metal-through bolts. Panel doors are made of 'structurally engineered' wood on the inside, with thin wood veneer on the outside to look like solid wood. This prevents warping, lessens the cost, and makes for a door that's as strong or stronger than solid wood.

In assessing your door, or shopping for a new one: Wood in-fill panels is the thinner wood used to give a panel door dimension and design quality. In-fill panels have molded edges, "tongues", that fit into grooves in the stiles and rails. Glass panels are the same way. Pay attention to tongue and groove thickness. The wider the tongues and grooves, the stronger the joinery. Stiles and rails with only 1/4" wide grooves can be kicked in; 1/2" thick grooves make a much stronger door. A door with narrow panels (vertical or horizontal) and thick tongues and grooves would require effort with a sledgehammer to break apart. (reminiscent of early castle doors!)

Flush wood doors may look like a solid plane of wood, but they aren't. The quality of materials and construction can vary greatly. Flush doors are comprised of three things: the outermost wooden frame for structure; the "core" or the space inside that is filled with materials that can be robust or cheap, (structurally engineered wood, such as Stave Core is good; foam or particleboard is bad); and visible on the outside, thinly sliced wood veneer. Flush doors are rated on quality and construction based on "wear"– the number of door slams per life of a door. They are *not tested* for resistance to battering or prying apart. The American Woodworking Institute (AWI) and the Wood Door Manufacturer's Association (WDMA) have the following standards based on wear: "Premium" grade is the best, most expensive. AWI's "Custom Grade" and WDMF's "Extra Heavy Duty or Heavy Duty" convey very good to adequate durability.

If you want the wood look and can afford the very best in security, some companies manufacture doors with over-the-top security features. Kreiger, for example, makes a steel framed, wood veneered door. www.kriegerproducts.com/kriegerwooddoors/

Fiberglass Doors

The finish options for fiberglass are impressive and attractive. Many emulate the wood look successfully. They have higher insulation values than wood. Many are hurricane rated–designed with interior metal reinforcing to withstand the impact of large objects thrown in high winds. This class of door is difficult to break open with force. Fiberglass doors come pre-hung in their own jamb and fit standard door size openings. Custom fabrication is very expensive. Modifications can be made however, to existing openings by altering the casing and jamb, which requires experienced carpentry and sizeable budget. Cheap fiberglass doors are to be avoided as they can have infill material similar to glued particleboard or foam, neither of which provide good purchase to screws. Good to high-end fiberglass tends to be relatively low-maintenance and durable. Door shields ("armoring") at the knob and deadbolt are recommended.

Steel Doors

For shear strength and impact resistance, steel doors can be the strongest option. Again, the quality of materials used and construction varies widely. Steel doors, like fiberglass, are prehung in their own steel jamb (unless a wood jamb is specified—but this will downgrade its overall strength). Like fiberglass, altering existing doorways gets expensive. Aesthetically, steel doors can emulate the wood look by 'pressing in' the panels. The end result can look inauthentic, or, nearly identical to wooden doors if they're well detailed and painted. The thickness of a door's steel determines its strength. A preferred gauge (thickness) is 14 (0.074"), which is considerably stronger than 18 gauge (0.047"). One manufacturer of 'steel' doors makes them with wood frames and covers them with paper thin panels of 25 gauge steel (.0215"). Don't buy those! At the core of a steel door are metal channels often filled with cardboard or plastic foam. Insulation values vary widely. Finishes are always opaque. One can buy a steel door that is already primed and ready to be painted. A very good, durable finish is a factory paint job over zinc-plated steel. Keep in mind: steel rusts more easily than wood rots. Be careful of exposure to humidity and moisture. Getting all the things you need out of a residential steel door can be a difficult search, and prices go up with features.

FOOTNOTES

INTRODUCTION

1. http://apps.pittsburghpa.gov/pghbop/2012_Annual_Report_v2.pdf

2. Ibid.

PART I

1. Glaeser, Edward L. "Why Is There More Crime in Cities?" Journal of Political Economy, vol. 107, no. S6, 1 Dec. 1999. pp. S225-S258 JSTOR, JSTOR.

2. Langton, Lynn, et al. "Victimizations Not Reported Victimizations Not Reported." Bureau of Justice Statistics (BJS), National Crime Victimization Survey, www.bjs.gov/index.cfm?ty=pbdetail&iid=4393.

3. Ibid.

4. Truman, Jennifer, and Lynn Langton. "Criminal Victimization, 2014." Bulletin, US Dept. of Justice, Bureau of Justice Statistics, 29 Sept. 2015, www.bjs.gov/content/pub/pdf/cv14.pdf.

5. Reaves, Brian A. Census of State and Local Law Enforcement Agencies, 2008. Bureau of Justice Statistics, July 2011, www.bjs.gov/content/pub/pdf/csllea08.pdf.

The last tally taken in 2008, included: 12,501 local police departments (including tribal police); 3,063 sheriffs' offices; 50 primary state law enforcement agencies; and, 1,733 special jurisdiction agencies (503 four-year public schools, 253 two-year public schools, health facilities, post offices, federal building, parks and recreation, forest fire, fish & wildlife, harbors, railroads, ports & airports, fraud, aron, tax evasion, narcotics, alcohol, gamin, and racing).

6. Simon, David. "The Wire's Final Season and the Story Everyone Missed." David Simon, davidsimon.com/the-wires-final-season-and-the-story-everyone-missed/, and, "In Baltimore, No One Left to Press the Police." David Simon, davidsimon.com/in-baltimore-no-one-left-to-press-the-police/.

David Simon, author of Amazon's TV show, The Wire, was a newspaper journalist with the Baltimore Sun for 12 years. The links above are to his blog, where he discusses real-life unreported crimes, and political influence on the perception versus reality of crime.

7. John A. Eterno, Arvind Verma & Eli B. Silverman. Police Manipulations of Crime Reporting: Insiders' Revelations, 2014. Justice Quarterly, DOI: 10.1080/07418825.2014.980838

8. Poston, Ben. "FBI Crime Reporting Audits Are Shallow and Infrequent." Journal Sentinel, 18 Aug. 2012, archive.jsonline.com/watchdog/watchdogreports/fbi-crimereporting-audits-are-shallow-infrequent-cg5uvel-166665516.html.

9. "Table 4." FBI, ucr.fbi.gov/crime-in-the-u.s/2013/preliminary-semiannual-uniform-crime-

report-january-june-2013/tables/Table_4_January_to_June_2012-2013_Offenses_Reported_to_Law_Enforcement_by_State_by_City.xls/view.

10. Sanchez, Ray. "William Bratton and Ray Kelly Spar over Crime Stats." CNN, Cable News Network, 31 Dec. 2015, www.cnn.com/2015/12/31/us/new-york-bratton-kelly-stats-feud/.

11. Langton, Lynn, and Marcus Berzofsky, Christopher Krebs, and Hope Smiley-McDonald. United States Department of Justice, Office of Justice Programs, Bureau of Justice Statistics. National Crime Victimization Survey. Table. 1. "Victimizations not reported to the police and the most important reason they went unreported, by the type of crime, 2006-2010. www.bjs.gov/content/pub/pdf/vnrp0610.pdf.

Total Crimes estimated to not be reported: 13,998,600. Crimes include: Violent: (Rape/sexual assault, Robbery, Aggravated assault), Simple assault; Personal larceny; Household Property: (Burglary, Motor vehicle theft, Theft). Reasons for not reporting include: Dealth with in another way/personal matter (20%); Not important enough to victim to victim to report (27%); Police would not or could not help (31%); Fear of reporisale or getting offender in trouble (5%); Other reason or not one most important reason (17%).

12. "Campus Safety and Security Data Analysis Cutting Tool ." Campus Safety and Security, US Dept. of Education, ope.ed.gov/security/index.aspx.

13. "Campus Sexual Violence: Statistics." Campus Sexual Violence: Statistics | RAINN, RAINN, www.rainn.org/statistics/campus-sexual-violence.

14. United States Dept. of Justice, Office of Justice Programs, Bureau of Justice Statistics. "Rape and Sexual Victimization Among College-Aged Females, 1995-2013.", 2014. www.bjs.gov/content/pub/pdf/rsavcaf9513.pdf.

15. "Department of Defense Annual Report on Sexual Assault in the Military." Sexual Assault Prevention and Response, Dept. of Defense, Fiscal Year 2012. www.sapr.mil/public/docs/reports/FY12_DoD_SAPRO_Annual_Report_on_Sexual_Assault-volume_one.pdf.

16. Kime, Patricia. "Sexual Assault Reporting Rises at U.S. Service Academies." Military Times, Sightline Media Group, 8 Aug. 2016. www.militarytimes.com/story/military/2016/01/08/reports-sexual-assaults-spike-military-academies/78496948/.

17. "McCain: Sex Assaults so Bad I Advise Women to Avoid Military." UPI, UPI, 5 June 2013, www.upi.com/McCain-Sex-assaults-so-bad-I-advise-women-to-avoid-military/90881370413800/.

On June 5, 2013. McCain delivered this testimony to the uniformed chiefs of the Army, Navy, Air Force, Marine Corps and Coast Guard at a Senate Armed Services Committee hearing.

18. "Burglary." FBI, FBI, 18 Aug. 2015, ucr.fbi.gov/crime-in-the-u.s/2014/crime-in-the-u.s.-2014/offenses-known-to-law-enforcement/burglary.

19 Truman, Jennifer L., and Lynn Langton. "Criminal Victimization, 2014." www.bjs.gov/content/pub/pdf/cv14.pdf.

The Bureau of Justice Statistics conducts the National Crime Victimization Survey. It uses national sampling, surveys and extensive follow ups, to understand the extent and nature of unreported crimes against people and property. Just as the FBI's Uniform Crime Report tallies reported crimes, the National Crime Victimization Survey estimates the much larger body of reported crimes.

20. DeMille, David. "Home Burglary Statistics: Will Your Home Be Broken Into?" ASecureLife.com, 2 Mar. 2017, www.asecurelife.com/burglary-statistics/.

21. "It's Taking Dallas Police Longer to Respond to 911 Calls Now, but Chief Says There's a Good Reason." Dallas News, The Dallas Morning News, 24 Aug. 2015, www.dallasnews.com/news/crime/2015/08/24/its-taking-dallas-police-longer-to-get-to-911-calls-now-but-chief-says-theres-a-good-reason.

22 "Average Police Emergency Response Time: about Ten Minutes." Exemplary Policing, Michael Josephson, 29 Aug. 2016, josephsonsexemplarypolicing.org/2016/08/average-response-time-about-ten-minutes/.

23. R., G. "In New Orleans, Call 911 and Wait for an Hour." Democracy in America, The Economist, 10 Dec. 2015, www.economist.com/blogs/democracyinamerica/2015/12/police-response-times.

24. "Is It Snoring or Sleep Apnea." SleepApnea.org, American Sleep Apnea Association, www.sleepapnea.org/learn/sleep-apnea/do-i-have-sleep-apnea/is-it-snoring-or-sleep-apnea/.

25. Main, Douglas. "30 Percent of Americans Have Had an Alcohol-Use Disorder." Tech & Science, Newsweek, 1 Apr. 2016, www.newsweek.com/30-percent-americans-have-had-alcohol-use-disorder-339085.

26. NationalSleepFoundation.com, sleepfoundation.org/bedroom/hear.php.

27. "CDC: 9 Million Americans Use Sleeping Pills." NY Daily News, 30 Aug. 2013, www.nydailynews.com/life-style/health/cdc-9-million-americans-sleeping-pills-article-1.1441778.

28. James, Steve. "1 In 8 Boomers Reports Memory Loss, Large Survey Finds." NBCNews.com, NBCUniversal News Group, 9 May 2013, www.nbcnews.com/health/1-8-boomers-reports-memory-loss-large-survey-finds-1C9864807.

PART 2

1. Levinson, David, ed. "Burglary." Encyclopedia of Crime and Punishment 1st Edition, Sage Publications, 2002.

2. Catalano, Shannan. "Victimization During Household Burglary". Special Report, US Dept. of Justice Statistics, September 2010, www.bjs.gov/content/pub/pdf/vdhb.pdf

PART 3

1. Kates, Dan, ed. Restricting Handguns–The Liberal Skeptics Speak Out. North River Press, 1979.

2. Eaton, S. Boyd, and Melvin Konner. "Paleolithic Nutrition." New England Journal of

Medicine, vol. 312, no. 5, 1985, pp. 283–289., doi:10.1056/nejm198501313120505.

3. "Adrenaline." You and Your Hormones, Society for Endocrinology, Jan. 2015, www.yourhormones.info/hormones/adrenaline.aspx.

4. Foreman, Carl, et al. High noon. 1952. A Stanley Kramer Production. A United Artists Release, (original).

5. Barruga, Aaron, and Andre M. Dall'au, et al. "Fearless Florida Senior Fights Intruder Using Frying Pan." Gun News and Gun Reviews, Tactical Life Gun Magazine, 18 Apr. 2011, www.tactical-life.com/news/fearless-florida-senior-fights-intruder-using-frying-pan/.

6. The Well Armed Woman, "Could You Use It?", http://thewellarmedwoman.com/could-you-use-it

7. Shooting the Bull, "Does Caliber Even Matter?", April 6, 2015. http://shootingthebull.net/blog/does-caliber-even-matter/

8. Dr Andreas Grabinsky, Anasthesiologist, on Gunshot Wounds. www.youtube.com/watch?v=wXwPtP-KDNk

Related to this (and expressed in later bullet points) is the idea that shooting someone does not equate to killing them immediately, or at all. Wounding, yes. In the following example, this established observation is expressed by Dr. Fackler, retired head of the Wound Ballistics Laboratory, US Army Medical Training Center, Letterman Institute, commenting on a man who survived being shot 20 times with a .22 caliber rifle at five feet, he said, "Shots to roughly 80 percent of targets on the body would not be fatal blows. It is like roulette". To consider shooting for home defense, one must think about the power of a gun, the potential velocity of the load used, shot placement, and the likelihood for multiple shots to stop a bad guy. Let alone accessing a firearm by day or night and being able to wield it effectively under pressure. If you plan to be a lethally armed home defender, train well and practice a lot. But foremost in this author's opinion: you want prevention more than self defense! A firearm is a tool of last resort.

BIBLIOGRAPHY

Applegate, Colonel Rex. *Kill or Get Killed*. Boulder: Paladin Press, 1976.

Applegate, Colonel Rex. Michael D. Janich. Bullseyes *Don't Shoot Back: The Complete Textbook of Point Shooting for Close Quarter Combat*. Boulder: Paladin Press, 1998.

Ayoob, Massad F. *In the Gravest Extreme*. Concord: Police Bookshelf, 1980.

Branca, Andrew. *The Law of Self Defense*. Maynard: XX, 2013

Castenada, Carlos. *Journey to Ixtlan*. New York: Simon & Schuster, 1972.

Chapman, Bob, Richter, Greg. *Realistic Self Defense for Ordinary People: A Course in Bulldog Jeet Kune Do*. Palatka: Bob Chapman, Greg Richter, 1996.

deBecker, Gavin. *The Gift of Fear – Survival Signals that Protect Us from Violence.* New York, Toronto: Little, Brown and Company, 1997.

Fairbairns, William Ewart. *Get Tough.* New York: D. Appleton-Century Company, 1942. Scribd. December 2014.

Fairbairns, William Ewart. *Hands Off! Self Defense for Women.* New York: D. Appleton-Century Company, 1942. Scribd. December 2014.

Gedgaudas, Nora. *Primal Body Primal Mind.* Rochester: Healing Arts Press, 2011.

Jones, Jan. *Self Defense Requires No Apologies.* Phoenix: Security World Publications, 1985.

Jordan, William H. *No Second Place Winner.* Shreveport: Wm. Jordan, 1980.

Kelly, D.O., Michael. *Death Touch: The Science Behind the Legend of Dim-Mak.* Boulder: Paladin Press, 2001.

Kruse M.D., Jack. *Epi-Paleo Rx: The Prescription for Disease Reversal and Optimal Health.* Optimized Life PLC, 2013.

MacLean, John. *Secrets of a Superthief.* New York: Berkley Trade, 1983.

Lovette, Ed, Dave Spaulding. *Defensive Living.* Flushing: Looseleaf Law Publications, 2005.

McGuff M.D., Doug and John Little. *Body by Science: A Research-based Program for Strength Training, Body Building, and Complete Fitness in 12 Minutes a Week.* New York: McGraw-Hill, 2009.

National Rifle Association. *NRA Guide to the Basics of Pistol Shooting.* Fairfax: NRA, 2009,

National Rifle Association. *NRA Guide to the Basics of Personal Protection in the Home.* Fairfax: NRA, 2000.

Renier, Noreen. *The Practical Psychic – A No-nonsense Guide to Developing Your Natural Intuitive Abilities.* Avon: Adam's Media, 2011.

Siddle, Bruce K. *Sharpening the Warrior's Edge.* Millstadt: PPCT Research Publications, 1995.

Soalt, Melissa with Paladin Press. *Fierce and Female.* Boulder: Paladin Press, 2000.

Tegner, Bruce. *Self Defense: Nerve Centers & Pressure Points for Karate, Jiujitsu, Atemi-Waza.* Ventura: Thor Publishing Co., 1987

VanCook, Jerry. *Real World Self Defense: A Guide to Staying Alive in Dangerous Times.* Boulder: Paladin Press, 1999.

Vilos, Mitch, Evan Vilos. *Self Defense Laws of All Fifty States.* Centerville: Guns West Publishing, 2013.

IMAGE CREDITS

All images from outside sources are noted.
All other images and illustrations are the property of Calamity May, LLC. Permission necessary to reprint.

p. 8   Sir Edward Coke, portrait during tenure as Chief Justice of the King's Bench, 1593. Artist Unknown. Accessed on Wikipedia.com

p. 44   Capt. Francis Hooke' Letter from John Farmer's The History of New Hampshire, 1831.

p. 48   Windsor Castle by Wenceslas Hollar, 17th century. Thomas Fisher Rare Book Library

Wikipedia.

p. 49   Windsor Castle photograph. Plan of Skenfrith Castle. St. J. O'Neill, Her Majesty's Stationery Office - Castles: An Introduction to the Castles of England and Wales. Accessed on Wikipedia.com

p. 55   Econo-Sign: 6.3"x9" - "Private Property / No Trespassing". Accessed at Amazon.com

p. 55   Dakota Alert Receiver www.dakotaalert.com/store/2500-series-products/dcr-2500-receiver/

p. 56   Wireless Driveway Alarm, https://www.dakotaalert.com/store/2500-series-products/dcpa-2500-driveway-alarm-transmitter/

p. 56   Wireless Motion Alert, https://www.dakotaalert.com/store/2500-series-products/dcmt-2500-passive-infrared-wireless-motion-detector/

p. 57   300-foot long Wireless Break Beam, https://www.dakotaalert.com/store/2500-series-products/00bbt-2500/

p. 57   Motion Sensor for Smaller Spaces, https://www.dakotaalert.com/store/2500-series-products/dcir-2500-pir-sensor/

p. 58   Motion Sensor and Digital Recorder, https://www.dakotaalert.com/store/video-products/dvr-01/

p. 58   Wireless Floor Mat, https://www.dakotaalert.com/store/1000-series-products/wps-1000-wireless-pull-station-clone/

p. 59   Digital Door Viewer, www.mul-t-lock-online.com

p. 59   290-degree Door Viewer Scope, http://new-vuetrading.com/store/index. html (bottom of page)

p. 59   Illustration, Photos courtesy of DavStar Safety & Security.

p. 74   Multi-Point Locking System, http://www.nicksbuilding.com/3_point_ locks/multipoint_locking_system.jpg

p. 74   EZ Armor Max Combo Set, White, http://armorconcepts.com/door-armor- max.html

p. 75   Door Movement Sensor, simplisafe.com, http://simplisafe.com/extra-en- try-sensor

p. 80   Entry Sensor, http://simplisafe.com/extra-entry-sensor

p. 80   Entry Sensor, small image, http://simplisafe.com/extra-entry-sensor

p. 80   Glass Break Sensor, http://simplisafe.com/glassbreak-sensor

p. 80   Receiver / Base Station, http://simplisafe.com

p. 81   Window Pin Lock, www.primeline.net/u-9857-window-pin-lock-18-x-1- 516-steel-chrome-plated

p. 102   Fairbairn Demonstrating Chin Jab & Imprint, Shanghai Municipal Police Self Defense Manual by W.E. Fairbairn (1915), accessed at Scribd.com

p. 102   Cover, William Fairbairn's Hands Off!, (1907), image accessed at Amazon. com

p. 122   "Jacksonville burglar gets hit 'upside the head' by 81-year-old victim's frying pan." Jacksonville Florida Times Union. April 13, 2011. Accessed at Jacksonville.com

p. 148   Winchester PDX1, .410, www.winchester.com/products/new-products/ pages/pdx1-410.aspx

p. 149   Mossberg 500 Tactical - HS410 Home Security, http://www.mossberg.com/ product/500-tactical-hs410-home-security-50359/

p. 150   Remington 11-87 Sportsman Compact, www.cabelas.com/product/Rem- ington-Model-Sportsman-Field-Semiautomatic-Shotgun/753150.uts